# Applied Mathematical Sciences

EDITORS

**Fritz John**
*Courant Institute of Mathematical Sciences*
New York University
New York, NY 10012

**J.E. Marsden**
*Department of Mathematics*
University of California
Berkeley, CA 94720

**Lawrence Sirovich**
*Division of Applied Mathematics*
Brown University
Providence, RI 02912

ADVISORS

**H. Cabannes** University of Paris-VI

**M. Ghil** New York University

**J.K. Hale** Brown University

**J. Keller** Stanford University

**G.B. Whitham** California Inst. of Technology

## EDITORIAL STATEMENT

The mathematization of all sciences, the fading of traditional scientific boundaries, the impact of computer technology, the growing importance of mathematical-computer modelling and the necessity of scientific planning all create the need both in education and research for books that are introductory to and abreast of these developments.

The purpose of this series is to provide such books, suitable for the user of mathematics, the mathematician interested in applications, and the student scientist. In particular, this series will provide an outlet for material less formally presented and more anticipatory of needs than finished texts or monographs, yet of immediate interest because of the novelty of its treatment of an application or of mathematics being applied or lying close to applications.

The aim of the series is, through rapid publication in an attractive but inexpensive format, to make material of current interest widely accessible. This implies the absence of excessive generality and abstraction, and unrealistic idealization, but with quality of exposition as a goal.

Many of the books will originate out of and will stimulate the development of new undergraduate and graduate courses in the applications of mathematics. Some of the books will present introductions to new areas of research, new applications and act as signposts for new directions in the mathematical sciences. This series will often serve as an intermediate stage of the publication of material which, through exposure here, will be further developed and refined. These will appear in conventional format and in hard cover.

## MANUSCRIPTS

The Editors welcome all inquiries regarding the submission of manuscripts for the series. Final preparation of all manuscripts will take place in the editorial offices of the series in the Division of Applied Mathematics, Brown University, Providence, Rhode Island.

SPRINGER-VERLAG NEW YORK INC., 175 Fifth Avenue, New York, N. Y. 10010

*Printed in U.S.A.*

# Applied Mathematical Sciences | Volume 55

# Applied Mathematical Sciences

1. John: **Partial Differential Equations**, 4th ed.
2. Sirovich: **Techniques of Asymptotic Analysis.**
3. Hale: **Theory of Functional Differential Equations**, 2nd ed.
4. Percus: **Combinatorial Methods.**
5. von Mises/Friedrichs: **Fluid Dynamics.**
6. Freiberger/Grenander: **A Short Course in Computational Probability and Statistics.**
7. Pipkin: **Lectures on Viscoelasticity Theory.**
8. Giacaglia: **Perturbation Methods in Non-Linear Systems.**
9. Friedrichs: **Spectral Theory of Operators in Hilbert Space.**
10. Stroud: **Numerical Quadrature and Solution of Ordinary Differential Equations.**
11. Wolovich: **Linear Multivariable Systems.**
12. Berkovitz: **Optimal Control Theory.**
13. Bluman/Cole: **Similarity Methods for Differential Equations.**
14. Yoshizawa: **Stability Theory and the Existence of Periodic Solutions and Almost Periodic Solutions.**
15. Braun: **Differential Equations and Their Applications**, 3rd ed.
16. Lefschetz: **Applications of Algebraic Topology.**
17. Collatz/Wetterling: **Optimization Problems.**
18. Grenander: **Pattern Synthesis: Lectures in Pattern Theory, Vol I.**
19. Marsden/McCracken: **The Hopf Bifurcation and its Applications.**
20. Driver: **Ordinary and Delay Differential Equations.**
21. Courant/Friedrichs: **Supersonic Flow and Shock Waves.**
22. Rouche/Habets/Laloy: **Stability Theory by Liapunov's Direct Method.**
23. Lamperti: **Stochastic Processes: A Survey of the Mathematical Theory.**
24. Grenander: **Pattern Analysis: Lectures in Pattern Theory, Vol. II.**
25. Davies: **Integral Transforms and Their Applications.**
26. Kushner/Clark: **Stochastic Approximation Methods for Constrained and Unconstrained Systems.**
27. de Boor: **A Practical Guide to Splines.**
28. Keilson: **Markov Chain Models—Rarity and Exponentiality.**
29. de Veubeke: **A Course in Elasticity.**
30. Sniatycki: **Geometric Quantization and Quantum Mechanics.**
31. Reid: **Sturmian Theory for Ordinary Differential Equations.**
32. Meis/Markowitz: **Numerical Solution of Partial Differential Equations.**
33. Grenander: **Regular Structures: Lectures in Pattern Theory, Vol. III.**
34. Kevorkian/Cole: **Perturbation Methods in Applied Mathematics.**
35. Carr: **Applications of Centre Manifold Theory.**

*(continued on inside back cover)*

K. Yosida

# Operational Calculus

## A Theory of Hyperfunctions

Springer-Verlag
New York Berlin Heidelberg Tokyo

K. Yosida
3-24-4 Kajiwara
Kamakura 247
Japan

AMS Classification: 33A99, 44A40, 44A45

Library of Congress Cataloging in Publication Data
Yoshida Kôsaku
Operational calculus.
(Applied mathematical sciences ; v. 55)
Bibliography: p.
Includes index.
1. Calculus, Operational. 2. Hyperfunctions. I. Title.
II. Series: Applied mathematical sciences (Springer-Verlag New York Inc.) ; v. 55.
QA432.Y6713 1984 515.7′2 84-10614

Title of the original Japanese edition: *Enzanshi-hō*, University of Tokyo Press, Tokyo, © 1982 by Kôsaku Yosida.



Printed and bound by R.R. Donnelley & Sons, Harrisonburg, Virginia.
Printed in the United States of America.

9 8 7 6 5 4 3 2 1

ISBN 0-387-96047-3 Springer-Verlag New York Berlin Heidelberg Tokyo
ISBN 3-540-96047-3 Springer-Verlag Berlin Heidelberg New York Tokyo

# Preface

In the end of the last century, Oliver Heaviside inaugurated an *operational calculus* in connection with his researches in electromagnetic theory. In his operational calculus, the *operator of differentiation* was denoted by the symbol "p". The explanation of this operator $p$ as given by him was difficult to understand and to use, and the range of the validity of his calculus remains unclear still now, although it was widely noticed that his calculus gives correct results in general.

In the 1930s, Gustav Doetsch and many other mathematicians began to strive for the mathematical foundation of Heaviside's operational calculus by virtue of the *Laplace transform*

$$\int_0^\infty e^{-pt} f(t)dt.$$

However, the use of such integrals naturally confronts restrictions concerning the growth behavior of the numerical function $f(t)$ as $t \to \infty$.

At about the midcentury, Jan Mikusiński invented the *theory of convolution quotients,* based upon the *Titchmarsh convolution theorem:*

> If $f(t)$ and $g(t)$ are continuous functions defined on $[0,\infty)$ such that the *convolution* $\int_0^t f(t-u)g(u)du \equiv 0$, then either $f(t) \equiv 0$ or $g(t) \equiv 0$ must hold.

The convolution quotients include the *operator of differentiation* "s" and related operators. Mikusiński's operational calculus gives a satisfactory basis of Heaviside's operational calculus; it can be applied successfully to *linear ordinary differential equations with constant coefficients* as well as to the *telegraph equation* which includes both the *wave and heat equations with constant coefficients*.

The aim of the present book is to give a simplified exposition as well as an extension of Mikusiński's operational calculus.

As for the *simplification*, I should like to mention two points 1° and 2° below.

1°. We give a plain proof of the Titchmarsh convolution theorem by making use of the well-known Liouville Theorem in *analytic function theory*.

2°. For solving linear ordinary differential equations with constant coefficients, we need not rely upon the Titchmarsh convolution theorem. We need only a rather trivial theorem:

> Let $f(t)$ be continuous for $0 \leqq t < \infty$ and $\int_0^t f(u)du \equiv 0$. Then $f(t) \equiv 0$.

As for the *extension*, I should like to mention the following point 3°.

3°. We define the *general power* $(s-\alpha)^\gamma$ of the operator $(s-\alpha)$ ($\alpha$ and $\gamma$ are complex numbers), by making use of the *general binomial theorem* in *analytic function theory:*

$$(1-z)^\gamma = \sum_{k=0}^{\infty} \binom{\gamma}{k}(-z)^k \quad \text{(convergent for } |z| < 1).$$

Then, by virtue of the general power $(s-\alpha)^\gamma$, we can *solve algebraically* the so-called *Laplace's differential equation:*

$$(a_2t+b_2)y''(t) + (a_1t+b_1)y'(t) + (a_0t+b_0)y(t) = 0,$$

where both the $a$'s and $b$'s are complex numbers and $a_2 \neq 0$. This equation includes Bessel's, Laguerre's and confluent hypergeometric differential equations and the like. It is to be noted here that Henri Poincaré and Emile Picard inaugurated the treatment of such differential equations by the Laplace transform combined with subtle contour integrations in the complex plane.

The present book is a revised and enlarged (by 1° and 3°) English Edition of the author's book Operational Calculus, written in Japanese and published by the University of Tokyo Press (1982). The English translation was done by the author.

The author gratefully acknowledges fine help from many friends; Hiroshi Fujita, Heinz Götze, Shûichi Okamoto, Shigetake Matsuura, Kyûya Masuda and Jan Mikusiński. Fujita kindly invited the author to write the Japanese Edition for the U.T.P. Götze of the Springer-Verlag kindly suggested the English edition. The above mentioned 1° and 2° were obtained by the author jointly with Matsuura and Okamoto, respectively. Masuda was a fine critic on $3^{\circ}$ during its maturity. Mikusiński's fine work aroused the author's interest towards the operational calculus. To them all, including the U.T.P. and the Springer-Verlag, Inc., I express my sincere thanks.

Kôsaku Yosida

Kamakura
July 1983

# Contents

Page

Preface v

Part I. INTEGRATION OPERATOR h AND DIFFERENTIATION OPERATOR s (CLASSES OF HYPERFUNCTIONS: $C$ AND $C_H$) 1

Chapter I. INTRODUCTION OF THE OPERATOR h THROUGH THE CONVOLUTION RING $C$ 1

1. Convolution Ring 1
2. Operator of Integration h 3

Chapter II. INTRODUCTION OF THE OPERATOR s THROUGH THE RING $C_H$ 5

3. The Ring $C_H$ and the Identity Operator $I = h/h$ 5
4. $C_H$ as a Class of Generalized Functions of Hyperfunctions 8
5. Operator of Differentiation s and Operator of Scalar Multiplication $[\alpha]$ 9
6. The Theorem $\frac{I}{s-[\alpha]} = e^{\alpha t}$ 12

Chapter III. LINEAR ORDINARY DIFFERENTIAL EQUATIONS WITH CONSTANT COEFFICIENTS 14

7. The Conversion of the Initial Value Problem for the Differential Equation into a Hyperfunction Equation 14
8. The Polynomial Ring of Polynomials in s has no Zero Factors 15
9. The Partial Fraction Decomposition of a Rational Function of s 18
10. Hyperfunction Solution of the Ordinary Differential Equation (The Operational Calculus) 22
11. Boundary Value Problems for Ordinary Differential Equations 29

Chapter IV. FRACTIONAL POWERS OF HYPERFUNCTIONS h, s AND $\frac{I}{s-\alpha}$ 32

12. Euler's Integrals - The Gamma Function and Beta Function 32
13. Fractional Powers of h, of $(s-\alpha)^{-1}$, and of $(s-\alpha)$ 34

Chapter V. HYPERFUNCTIONS REPRESENTED BY INFINITE POWER SERIES IN h 39

14. The Binomial Theorem 39
15. Bessel's Function $J_n(t)$ 40

Page

CHAPTER V. (cont.)

16. Hyperfunctions Represented by Power Series in h 42

PART II. LINEAR ORDINARY DIFFERENTIAL EQUATIONS WITH LINEAR COEFFICIENTS (THE CLASS C/C OF HYPERFUNCTIONS) 47

CHAPTER VI. THE TITCHMARSH CONVOLUTION THEOREM AND THE CLASS C/C 47

17. Proof of the Titchmarsh Convolution Theorem 47
18. The Class C/C of Hyperfunctions 50

CHAPTER VII. THE ALGEBRAIC DERIVATIVE APPLIED TO LAPLACE'S DIFFERENTIAL EQUATION 53

19. The Algebraic Derivative 53
20. Laplace's Differential Equation 60
21. Supplements. I: Weierstrass' Polynomial Approximation Theorem. II: Mikusiński's Theorem of Moments 70

PART III. SHIFT OPERATOR $\exp(-\lambda s)$ AND DIFFUSION OPERATOR $\exp(-\lambda s^{1/2})$ 74

CHAPTER VIII. EXPONENTIAL HYPERFUNCTIONS $\exp(-\lambda s)$ AND $\exp(-\lambda s^{1/2})$ 74

22. Shift Operator $\exp(-\lambda s) = e^{-\lambda s}$. Function Space $K = K[0,\infty)$ 74
23. Hyperfunction-Valued Function $f(\lambda)$ and Generalized Derivative $\frac{d}{d\lambda} f(\lambda) = f'(\lambda)$ 79
24. Exponential Hyperfunction $\exp(\lambda s) = e^{\lambda s}$ 83
25. Examples of Generalized Limit. Power Series in $e^{\lambda s}$ 86
26. $\int_0^\infty e^{-\lambda s} f(\lambda) d\lambda = \{f(t)\}$ For $\{f(t)\} \in C$ 92
27. Exponential Hyperfunction $\exp(-\lambda s^{1/2}) = e^{-\lambda s^{1/2}}$ 94
28. Logarithmic Hyperfunction w and Exponential Hyperfunction $\exp(\lambda w)$ 99

PART IV. APPLICATIONS TO PARTIAL DIFFERENTIAL EQUATIONS 106

CHAPTER IX. ONE DIMENSIONAL WAVE EQUATION 108

29. Hyperfunction Equation of the Form $f''(\lambda) - w^2 f(\lambda) = g(\lambda)$, $w \in C/C$ 108
30. The Vibration of a String 113
31. D'Alembert's Method 118
32. The Vibration of an Infinitely Long String 122

CHAPTER X. TELEGRAPH EQUATION 124

33. The Hyperfunction Equation of the Telegraph Equation 124
34. A Cable With Infinitely Small Loss 125

Page

CHAPTER X. (cont.)

35. Conductance Without Deformation 126
36. The Thomson Cable 128
37. Concrete Representations of $\exp(-\lambda\sqrt{\alpha s + \beta})$ 132
38. A Cable Without Self-Induction 138
39. A Cable Without Leak-Conductance 140
40. The Case Where All the Four Parameters Are Positive 143

CHAPTER XI. HEAT EQUATION 145

41. The Temperature of a Heat-Conducting Bar 145
42. An Infinitely Long Bar 147
43. A Bar Without an Outgoing Flow of Heat 149
44. The Temperature in a Bar with a Given Initial Temperature 150
45. A Heat-Conducting Ring 153
46. Non-Insulated Heat Conduction 156

ANSWERS TO EXERCISES 157

FORMULAS AND TABLES 160

REFERENCES 166

PROPOSITIONS AND THEOREMS IN SECTIONS 167

INDEX 169

# Part I
# Integration Operator h and Differentiation Operator s (Classes of Hyperfunctions: C and $C_H$)

# Chapter I
# Introduction of the Operator h Through the Convolution Ring C

## §1. CONVOLUTION RING

The totality of complex-valued continuous functions $a(t)$, $b(t)$, $f(t)$ and so forth defined on the interval $[0,\infty)$ will play a particularly important role in the *operational calculus*; we shall denote the class of those functions by $C[0,\infty)$ or simply by the letter $C$. The convolution of two functions $a = a(t)$ and $b = b(t)$ of $C$ is defined by

$$(a*b)(t) = a*b(t) = \int_0^t a(t-u)b(u)du \qquad (0 \leqq t < \infty), \tag{1.1}$$

and we have

PROPOSITION 1. $a*b$ belongs to $C$; i.e., $a*b(t)$ is a continuous function defined on $[0,\infty)$.

PROOF: Let $\delta$ be any positive number. We shall prove

$$\lim_{\delta\to 0} a*b(t+\delta) = a*b(t).$$

In fact, we have

$$\int_0^{t+\delta} a(t+\delta-u)b(u)du = \int_0^t a(t+\delta-u)b(u)du + \int_t^{t+\delta} a(t+\delta-u)b(u)du$$
$$= I_{1,\delta} + I_{2,\delta}.$$

For fixed $t > 0$, it is clear that

$$\lim_{\delta\to+0} I_{1,\delta} = \int_0^t a(t-u)b(u)du = a*b(u).$$

Also, for fixed $t > 0$ and $0 < \delta < 1$, we have

$$|I_{2,\delta}| \leqq \int_t^{t+\delta} AB du \leqq \delta \cdot AB, \text{ where}$$

$$A = \max_{0 \leqq s \leqq t+1} |a(s)|, \quad B = \max_{t \leqq u \leqq t+1} |b(u)|.$$

This proves $\lim_{\delta \to +0} I_{2,\delta} = 0$, and so we have obtained

$$\lim_{\delta \to +0} a*b(t+\delta) = a*b(t) \quad (\text{for} \quad t \geqq 0).$$

Similarly, we can prove $\lim_{\delta \to +0} a*b(t-\delta) = a*b(t)$ for fixed $t > 0$.

In the following, we shall denote a function in $C$ by $\{f(t)\}$ or simply by $f$, while $f(t)$ means the value at $t$ of this function $f$. And for the convolution $a*b$ of two functions $a$ and $b$ of $C$, we simply write $ab$ so that

$$ab(t) = \int_0^t a(t-u)b(u)du \quad (0 \leqq t < \infty). \tag{1.1'}$$

For $a,b \in C$, the *sum* $a+b$ is defined by

$$(a+b)(t) = a(t) + b(t) \tag{1.2}$$

and $(a+b) \in C$. For any $\alpha \in C^1$ (the field of complex numbers) and $a \in C$, the *scalar multiple* $\alpha a$ is defined by

$$(\alpha a)(t) = \alpha a(t). \tag{1.3}$$

We have the following:

THEOREM 1.

$$a+b = b+a, \quad (a+b)+c = a+(b+c), \tag{1.4}$$

$$a(b+c) = ab+ac, \quad \alpha(a+b) = \alpha a + \alpha b, \quad (\alpha+\beta)a = \alpha a + \beta a, \tag{1.5}$$

(Greek letters denote complex numbers.)

$$\alpha(ab) = (\alpha a)b = a(\alpha b), \quad (\alpha\beta)a = \alpha(\beta a), \tag{1.6}$$

$$ab = ba, \quad a(bc) = (ab)c, \quad (a+b)c = ac + bc. \tag{1.7}$$

PROOF: We shall prove (1.7) only, since the rest are easy. By the substitution $t-u = r$, we obtain

$$(ab)(t) = \int_0^t a(t-u)b(u)du = -\int_t^0 a(r)b(t-r)dr$$

$$= \int_0^t b(t-r)a(r)dr = (ba)(t).$$

Next, by using $a(bc) = (bc)a$ and changing the order of integration of the iterated integral, we obtain

$$((a(bc))(t) = ((bc)a)(t) = \int_0^t \left[\int_0^{t-u} b(t-u-v)c(v)dv\right] a(u)du$$

$$= \int_0^t \left[\int_0^{t-v} a(u)b(t-v-u)du\right] c(v)dv = ((ab)c)(t).$$

REMARK 1.1. The content of Theorem 1 is expressed in the words of *algebra* as follows: $C$ is a *ring* with respect to *addition* and the *multiplication* (or the *product*) $ab$, and the multiplication is *commutative*, i.e., $ab = ba$. The *zero* of this ring $C$ is the function $f \in C$ with $f(t) \equiv 0$ on $[0,\infty)$; we denote it by $0$ and

$$f+0 = f \quad \text{and} \quad f0 = 0. \tag{1.8}$$

EXAMPLES OF CONVOLUTION

EXAMPLE 1.1. For $a(t) = t$ and $b(t) = e^t$,

$$ab(t) = \int_0^t (t-u)e^u du = t\int_0^t e^u du - \int_0^t ue^u du$$

$$= t(e^t-1) - ue^u\Big|_{u=0}^{u=t} + \int_0^t e^u du$$

$$= t(e^t-1) - te^t + e^t - 1$$

$$= -t + e^t - 1.$$

EXAMPLE 1.2. For $a(t) = 1$ and $b(t) = \sin^2 t$,

$$\{1\}\{\sin^2 t\} = \int_0^t \sin^2 u du = -\sin u \cos u\Big|_{u=0}^{u=t}$$

$$+ \int_0^t \cos^2 u du = -\sin t \cos t + t - \int_0^t \sin^2 u du.$$

$$= \frac{1}{2}t - \frac{1}{2}\sin t \cos t.$$

## §2. OPERATOR OF INTEGRATION h

We shall denote the function $\{1\}$ by $h$. Then, as in Example 1.2, we have

$$(hf)(t) = \int_0^t f(u)du \qquad (f \in C), \tag{2.1}$$

so that the convolution product $hf$ is equal to the indefinite integral of f from 0 to t. Hence h represents the *operator of integration*.

REMARK 2.1. The introduction of the letter "h" is due to the historical fact that Oliver Heaviside (1850-1925), the founder of the *operational calculus*, used

$$H(t) = \begin{cases} 1, & t > 0 \\ 1/2, & t = 0 \\ 0, & t < 0. \end{cases} \tag{2.2}$$

for the "unit function" in his calculus. Jan Mikusiński used the letter "$\ell$" for 1 in his mathematical foundation of Heaviside's operational calculus. However, because of typographical reasons, the present book follows the use of "h" due to Erdelyi [3].

PROPOSITION 2.

$$h^2 = hh = \left\{\int_0^t du\right\} = \{t\}, \quad h^3 = hh^2 = \left\{\int_0^t udu\right\} = \{\frac{t^2}{2!}\},$$

and by induction,

$$h^n = \left\{\frac{t^{n-1}}{(n-1)!}\right\} \quad (n = 1,2,\ldots). \tag{2.3}$$

EXAMPLES. Verify the following equalities:

(α) $\{t\}^2 = \{t\}\{t\} = \{\frac{1}{6} t^3\} = (h^2)^2 = h^4$,

(β) $\{t\}^3 = \{t\}\{t\}^2 = \{\frac{1}{120} t^5\} = (h^2)^3 = h^6$,

(γ) $\{e^t\}^2 = \{e^t\}\{e^t\} = \{te^t\}$,

(δ) $\{e^t\}^3 = \{e^t\}\{e^t\}^2 = \{\frac{1}{2!} t^2 e^t\}$.

EXERCISE 2. Simplify the following expressions:

(α) $h\{\cos^2 t\} + h\{\sin^2 t\}$,

(β) $h^2\{\cos^2 t\} + h^2\{\sin^2 t\}$,

(γ) $\{1-t\}\{e^t\} + \{e^{-t}\}\{1+t\} - \{1-t\}\{e^{-t}\} - \{e^t\}\{1+t\}$,

(δ) $\{1-t^{1/2}\}\{\sin t\} + \{1+t^{1/2}\}\{\cos t\} + \{1-t^{1/2}\}\{\cos t\} + \{1+t^{1/2}\}\{\sin t\}$.

# Chapter II
# Introduction of the Operator s Through the Ring $C_H$

## §3. THE RING $C_H$ AND THE IDENTITY OPERATOR $I = \frac{h}{h}$

PROPOSITION 3. Let

$$H = \{k: k = h^n \ (n = 1,2,\ldots)\}.$$

Then, for any $k = h^n \in H$ and $f \in C$, we have

$$kf = 0 \text{ implies } f = 0, \text{ where } 0 = \{0\} \in C. \tag{3.1}$$

PROOF: The equation

$$hf(t) = \int_0^t f(u)\,du = 0 \qquad (0 \leqq t < \infty)$$

implies, by differentiation with respect to $t$, that

$$\frac{d}{dt}(hf)(t) = f(t) = 0 \qquad (0 \leqq t < \infty),$$

i.e., $f = 0$. Hence

$$h^2 f = h(hf) = 0 \text{ implies } hf = 0 \text{ and so } f = 0.$$

Therefore, by induction on $n$, we obtain (3.1).

As a corollary of Proposition 3, we obtain

PROPOSITION 4. Let $k, k' \in H$ and $f, f' \in C$. Then the following relation $\approx$ of two "fractions" $\frac{f}{k}$ and $\frac{f'}{k'}$, given by

$$\frac{f}{k} \approx \frac{f'}{k'} \text{ if and only if } k'f = kf', \tag{3.2}$$

is an *equivalence relation*. That is, we have

$$\frac{f}{k} \approx \frac{f}{k}, \tag{3.2)$_1$}$$

$$\frac{f}{k} \approx \frac{f'}{k'} \text{ implies } \frac{f'}{k'} \approx \frac{f}{k}, \qquad (3.2)_2$$

$$\frac{f}{k} \approx \frac{f'}{k'}, \quad \frac{f'}{k'} \approx \frac{f''}{k''} \text{ implies } \frac{f}{k} \approx \frac{f''}{k''} . \qquad (3.2)_3$$

PROOF: $(3.2)_1$ is evident from $kf = kf$. $(3.2)_2$ is also evident since $k'f = kf'$ implies $kf' = k'f$ and vice versa.

PROOF OF $(3.2)_3$: We have $k'f = kf'$ and $k''f' = k'f''$ by assumption. Hence $k''k'f = k''kf'$ and $kk''f' = kk'f''$ and so, by (1.7), we have

$$k'k''f = k''k'f = k''kf' = kk'f'' = k'kf''.$$

Thus

$$k'(k''f - kf'') = 0 \text{ which implies } k''f - kf'' = 0$$

by Proposition 3. The last equation means $\frac{f}{k} = \frac{f''}{k''}$.

REMARK 3.1. Since the equivalence relation holds, we can replace the above $\approx$ by the equality $=$ so that

$$\frac{f}{k} = \frac{f'}{k'} \text{ if and only if } k'f = kf'. \qquad (3.2)'$$

THEOREM 2. We can define the sum and product of two "fractions" $\frac{f_1}{k_1}$ and $\frac{f_2}{k_2}$ as follows.

$$\frac{f_1}{k_1} + \frac{f_2}{k_2} = \frac{f_1k_2 + f_2k_1}{k_1k_2}, \qquad (3.3)$$

$$\frac{f_1}{k_1}\frac{f_2}{k_2} = \frac{f_1f_2}{k_1k_2}. \qquad (3.4)$$

PROOF: We have to show the following: If $\frac{f_1}{k_1} = \frac{f_3}{k_3}$ and $\frac{f_2}{k_2} = \frac{f_4}{k_4}$, then

$$\frac{f_1}{k_1} + \frac{f_2}{k_2} = \frac{f_3}{k_3} + \frac{f_4}{k_4}, \quad \frac{f_1}{k_1}\frac{f_2}{k_2} = \frac{f_3}{k_3}\frac{f_4}{k_4}$$

hold. That is, we have to prove that

$$\frac{f_1k_2 + f_2k_1}{k_1k_2} = \frac{f_3k_4 + f_4k_3}{k_3k_4}, \quad \frac{f_1f_2}{k_1k_2} = \frac{f_3f_4}{k_3k_4}$$

follow from

$$k_3f_1 = k_1f_3 \quad \text{and} \quad k_4f_2 = k_2f_4.$$

This can be shown as in the case of ordinary numerical fractions.

COROLLARY. By (3.3) and (3.4), the set

$$C_H = \{\frac{f}{k} : f \in C \text{ and } k \in H\}^* \tag{3.5}$$

is a ring whose "product" is *commutative*. Namely, the following $(3.5)_1$-$(3.5)_4$ hold:

$$\begin{cases} \frac{f}{k} + \frac{f'}{k'} = \frac{f'}{k'} + \frac{f}{k}, \\ \frac{f}{k} + \left(\frac{f'}{k'} + \frac{f''}{k''}\right) = \left(\frac{f}{k} + \frac{f'}{k'}\right) + \frac{f''}{k''}. \end{cases} \tag*{$(3.5)_1$}$$

$$\begin{cases} \frac{f}{k}\left(\frac{f'}{k'} + \frac{f''}{k''}\right) = \frac{f}{k}\frac{f'}{k'} + \frac{f}{k}\frac{f''}{k''}, \\ \alpha\left(\frac{f}{k} + \frac{f'}{k'}\right) = \frac{\alpha f}{k} + \frac{\alpha f'}{k'}, \\ (\alpha+\beta)\frac{f}{k} = \frac{\alpha f}{k} + \frac{\beta f}{k}. \end{cases} \tag*{$(3.5)_2$}$$

$$\begin{cases} \alpha\left(\frac{f}{k}\frac{f'}{k'}\right) = \frac{\alpha f}{k}\frac{f'}{k'} = \frac{f}{k}\frac{\alpha f'}{k'}, \\ (\alpha\beta)\frac{f}{k} = \alpha\left(\beta\frac{f}{k}\right) = \alpha\left(\frac{\beta f}{k}\right) = \frac{(\alpha\beta) f}{k}. \end{cases} \tag*{$(3.5)_3$}$$

$$\begin{cases} \frac{f}{k}\frac{f'}{k'} = \frac{f'}{k'}\frac{f}{k}, \quad \frac{f}{k}\left(\frac{f'}{k'}\frac{f''}{k''}\right) = \left(\frac{f}{k}\frac{f'}{k'}\right)\frac{f''}{k''}, \\ \left(\frac{f}{k} + \frac{f'}{k'}\right)\frac{f''}{k''} = \frac{f}{k}\frac{f''}{k''} + \frac{f'}{k'}\frac{f''}{k''}. \end{cases} \tag*{$(3.5)_4$}$$

The proof is obtained as in the case of numerical fractions.

THE IDENTITY OPERATOR I. By (3.2)', we have

$$\frac{h}{h} = \frac{h^1}{h^1} = \frac{h^2}{h^2} = \ldots = \frac{h^n}{h^n} = \ldots$$

Also we have, by $kh^n f = h^n kf$,

$$\frac{h^n}{h^n}\frac{f}{k} = \frac{f}{k} \quad (k \in H). \tag{3.6}$$

Therefore, $\frac{h^n}{h^n}$ is the *multiplicative unit* of the ring $C_H$. Hence we shall denote $\frac{h^n}{h^n}$ by I or simply by 1 and call I the *unit (or identity) operator:*

---

*Yosida, K. and Okamoto, S.: A note on Mikusiński's operational calculus, Proc. Jap Acad., *56*, Ser. A, No. 1 (1980), 1-3.

$$\frac{h^n}{h^n} = I = 1 \quad (n = 1,2,\ldots). \tag{3.7}$$

## §4. $C_H$ AS A CLASS OF GENERALIZED FUNCTIONS OR HYPERFUNCTIONS

The ring $C$ is a *subring* of the ring $C_H$, since we can identify $f = \{f(t)\} \in C$ with $\frac{h^n f}{h^n} \in C_H$, i.e.,

$$f = \frac{h^n f}{h^n} \quad (n = 1,2,\ldots). \tag{4.1}$$

In fact, by (3.3)-(3.4), we have

$$\frac{h^n f}{h^n} + \frac{h^m g}{h^m} = \frac{h^{n+m} f + h^{n+m} g}{h^{n+m}} = \frac{h^{n+m}(f+g)}{h^{n+m}}$$

and

$$\frac{h^n f}{h^n}\,\frac{h^m g}{h^m} = \frac{h^{n+m}(fg)}{h^{n+m}} .$$

Therefore, $C_H$ contains $\frac{h^n f}{h^n} = f$, which is a continuous function. However, $C_H$ is actually bigger than $C$ as the following proposition shows.

PROPOSITION 5. The unit operator $I$ does not belong to $C$.

PROOF: We shall derive an absurdity from $I = \frac{h}{h} = f$ where $f \in C$. In fact, from $\frac{h}{h} = \frac{hf}{h}$, we obtain $h^2 = h^2 f$, so that

$$\{t\} = \{t\}f = \int_0^t (t-u)f(u)du.$$

Hence, by differentiation, we obtain

$$\{1\} = \left\{(t-t)f(t) + \int_0^t f(u)du\right\} = \left\{\int_0^t f(u)du\right\}.$$

Thus, by further differentiation, we have

$$0 = \{f(t)\}, \quad \text{i.e.,} \quad f(t) = 0 \quad \text{for all} \quad t \geqq 0.$$

This yields a contradiction: $\{t\} = h^2 = h^2 f = h^2 0 = 0$.

GENERALIZED FUNCTIONS OR HYPERFUNCTIONS. The equality $I = \frac{h}{h}$ shows that from a continuous function $h \in C$ we can obtain an element $\frac{h}{h} \in C_H$ which is not a continuous function. Hence $I$ is a *generalized function* or, preferably*, a *hyperfunction***, obtained by the process of "dividing" elements of $C$ by $h$. We shall further obtain, in §5, another hyperfunction $s$ which is the *operator of differentiation.*

## §5. OPERATOR OF DIFFERENTIATION s AND OPERATOR OF SCALAR MULTIPLICATION [α]

THEOREM 3. We put

$$s = \frac{1}{h} = \frac{I}{h} = \frac{h^{n-1}}{h^n} \qquad (n = 1,2,\ldots;\ h^0 = I). \tag{5.1}$$

$s$ is the *operator of differentiation* in the following sense: If both $f$ and its derivative $f'$ belong to $C$, then

$$f' = sf - [f(0)], \quad \text{where} \quad [f(0)] = s\{f(0)\} = \frac{\{f(0)\}}{h}, \tag{5.2}$$

so that both $sf$ and $[f(0)]$ belong to $C_H$.

PROOF: We have

$$sh = hs = h\frac{I}{h} = I = 1. \tag{5.3}$$

Hence, by the fundamental theorem of calculus, we have

$$\begin{aligned} hf' &= \{(hf')(t)\} = \left\{\int_0^t f'(u)du\right\} = \{f(t) - f(0)\} \\ &= \{f(t)\} - \{f(0)\} = \{f(t)\} - h[f(0)], \end{aligned} \tag{5.3)'$$

and so, by (5.3),

$$shf' = (sh)f' = f' = sf - sh[f(0)] = sf - [f(0)].$$

THEOREM 4. We denote, for any complex number $\alpha$,

$$[\alpha] = \frac{\{\alpha\}}{h} \in C_H. \tag{5.4}$$

Then $[\alpha]$ is the operator of *scalar multiplication* in the following sense:

$$\text{For any } \frac{f}{h^n} \in C_H, \text{ we have } [\alpha]\frac{f}{h^n} = \frac{\alpha f}{h^n}. \tag{5.5}$$

---

*The adjective "generalized" is somewhat lengthy.

**Although this term "hyperfunction" has been used since 1959 by Professor Mikio Sato for his generalization of the notion of functions, there should be no serious confusion with my use due to the fact that Sato's generalization is based upon the (generalized) *boundary value of analytic functions*.

PROOF:

$$[\alpha]\frac{f}{h^n} = \frac{\{\alpha\}}{h}\frac{f}{h^n} = \frac{\{\alpha\}f}{h^{n+1}} = \frac{\{\int_0^t \alpha f(u)du\}}{h^{n+1}} = \frac{h(\alpha f)}{hh^n} = \frac{\alpha f}{h^n}.$$

This means that $[\alpha]$ acts as a scalar multiplication by $\alpha$ in the ring $C_H$.

REMARK 5.1. Defining $s^o = I$, we have

$$C_H = \{\frac{f}{h^n} : f \in C, \quad (n = 0,1,\ldots)\} = \{s^n f: f \in C, \quad (n = 0,1,\ldots)\}.$$

COROLLARY OF THEOREM 3. If $f, f', f'', \ldots, f^{(n)} \in C$, then

$$f^{(n)} = s^n f - s^{n-1}[f(0)] - s^{n-2}[f'(0)] - \ldots - s[f^{(n-2)}(0)] - [f^{(n-1)}(0)]. \tag{5.6}$$

PROOF: By (5.2), we have

$$f'' = sf' - [f'(0)] = s(sf - [f(0)]) - [f'(0)]$$
$$= s^2 f - s[f(0)] - [f'(0)].$$

We have already established (5.6) for $n = 1$. Hence the general case of (5.6) is proved by induction on $n$.

REMARK 5.2. When $f \in C = C[0,\infty)$ does not have its derivative $f'$ belonging to $C$, we shall call $sf = \frac{f}{h} \in C$ as the "generalized derivative" of $f$. Thus, in this sense, $f$ is *infinitely differentiable in a generalized sense* so that the *n-th generalized derivative of* $f$ is given by

$$s^n f = \frac{f}{h^n} \in C.$$

Thus $s^n f$ is a *hyperfunction* $(n = 1,2,\ldots)$, and

$$s^m(s^n f) = s^{n+m} f.$$

Further,

$$[\alpha] = \frac{\{\alpha\}}{h} = s\{\alpha\} = \alpha I$$

is also a *hyperfunction*. In the following, we shall write $[\alpha]$, $[f(0)]$, $[f'(0)]$, etc. simply by $\alpha$, $f(0)$, $f'(0)$ etc., in cases where no ambiguity will result.

REMARK 5.3. Any polynomial in $s$ given by

$$p(s) = \alpha_n s^n + \alpha_{n-1}s^{n-1} + \ldots + \alpha_1 s + \alpha_0 I \quad (I = s^o) \tag{5.7}$$

belongs to $C_H$, since $C_H$ is a ring and

$$\alpha_n s^n \in C_H \quad (n = 0,1,\dots).$$

If

$$p(s) = \alpha_n s^n + \alpha_{n-1}s^{n-1} + \dots + \alpha_0 I$$
$$= \beta_m s^m + \beta_{m-1}s^{m-1} + \dots + \beta_0 I \qquad (\text{with } n \geqq m),$$

then

$$\begin{cases} \alpha_n = \alpha_{n-1} = \dots = \alpha_{m+1} = 0 \text{ and} \\ \alpha_m = \beta_m,\ \alpha_{m-1} = \beta_{m-1}, \dots, \alpha_0 = \beta_0. \end{cases} \tag{5.8}$$

PROOF: In the case $n > m$, we must have

$$p(s) - p(s) = 0 = \alpha_n s^n + \alpha_{n-1}s^{n-1} + \dots + \alpha_{m+1}s^{m+1}$$
$$+ (\alpha_m - \beta_m)s^m + (\alpha_{m-1} - \beta_{m-1})s^{m-1} + \dots + (\alpha_0 - \beta_0)I,$$

and hence, by multiplying by $h^n$, we obtain

$$0 = \alpha_n h^n s^n + \dots + \alpha_{m+1}h^n s^{m+1} + (\alpha_m - \beta_m)h^n s^m + \dots + (\alpha_0 - \beta_0)h^n$$
$$= \alpha_n I + (\alpha_{n-1}h + \dots + \alpha_{m+1}h^{n-m-1} + (\alpha_m - \beta_m)h^{n-m} + \dots + (\alpha_0 - \beta_0)h^n$$
$$= \alpha_n I + \left(\alpha_{n-1}\{1\} + \dots + \alpha_{m+1}\left\{\frac{t^{n-m-2}}{(n-m-2)!}\right\} + (\alpha_m - \beta_m)\left\{\frac{t^{n-m-1}}{(n-m-1)!}\right\}\right.$$
$$\left. + \dots + (\alpha_0 - \beta_0)\left\{\frac{t^{n-1}}{(n-1)!}\right\}\right).$$

Therefore, by Proposition 5, we must have

$$\alpha_n I = 0 \quad \text{and} \quad \{\text{a polynomial in } t\} = 0.$$

Hence (5.8) is proved in the case $n > m$. The case $n = m$ is easily proved as above.

REMARK 5.4. In his book Operational Calculus, J. Mikusiński [5] introduced the "fraction" of the form

$$\frac{f}{g} \quad (f,g \in C \text{ and } g \neq 0)$$

which is more general than our "fraction"

$$\frac{f}{h^n} = s^n f \quad (f \in C).$$

Mikusiński's standpoint was based upon the so-called *Titchmarsh convolution theorem* which reads: If $a,b \in C$ be such that $ab = 0$, then either $a = 0$ or $b = 0$. By virtue of this theorem, we can define the "fraction" or the *convolution quotient*

$$\frac{f}{g} \quad (f,g \in C \text{ and } g \neq 0)$$

similarly to our case of §3:

$$\frac{f}{h^n} \quad (f \in C \text{ and } n = 1,2,\ldots).$$

Titchmarsh discovered the above theorem in 1926 by making use of deep theorems on analytic functions.* An elementary proof was given in 1951 by Mikusiśki jointly with C. Ryll-Nardzewski.** It is elementary in the sense that it does not appeal to analytic function theory, although it takes about 9 pages. A plain proof*** will be given in §17 of the present book by relying upon Liouville's theorem in analytic function theory. It is to be noted here that before Chapter VI of the present book, we do not need the Titchmarsh convolution theorem; we need only hyperfunctions of the form

$$s^n f \quad (n = 1,2,\ldots)$$

which were introduced by virtue of Proposition 3.

## §6. THE THEOREM $\frac{I}{s-[\alpha]} = \{e^{\alpha t}\}$

THEOREM 5. For any complex number $\alpha$, there exists one and only one solution $\frac{f}{h^n}$ of the equation

$$(s-[\alpha])\frac{f}{h^n} = I \tag{6.1}$$

given by $\frac{f}{h^n} = \{e^{\alpha t}\} \in C$.

PROOF: We have, by (5.2),

$$s\{e^{\alpha t}\} = \{\frac{d}{dt} e^{\alpha t}\} + [e^{\alpha \cdot 0}] = [\alpha]\{e^{\alpha t}\} + [1]$$

so that

---

*Edward Charles Titchmarsh (1899-1963): Introduction to the Theory of Fourier Integrals, Oxford University Press (1948), p. 327-.

**J. Mikusiński [5], p. 15 - p. 23.

***Kôsaku Yosida-Shigetake Matsuura: A Note on Mikusiński's Proof of the Titchmarsh Convolution Theorem, to be published in the Contemporary Mathematics series of the Amer. Math. Society.

$$(s-[\alpha])\{\dot{e}^{\alpha t}\} = [\alpha]\{e^{\alpha t}\} + [1] - [\alpha]\{e^{\alpha t}\} = [1] = I,$$
$$\{e^{\alpha t}\}(s-[\alpha]) = I \tag{6.2}$$

by the commutativity $(3.5)_4$ of the product in the ring $C_H$.

The uniqueness of the solution $\frac{f}{h^n}$ of (6.1) is proved as follows. By multiplying on both sides of (6.1) by $\{e^{\alpha t}\}$, we obtain

$$\{e^{\alpha t}\}((s-[\alpha])\frac{f}{h^n}) = \{e^{\alpha t}\}.$$

Hence, by the associativity $(3.5)_4$ of the product in $C_H$,

$$(\{e^{\alpha t}\}(s-[\alpha]))\frac{f}{h^n} = I\,\frac{f}{h^n} = \{e^{\alpha t}\},$$

that is, $\frac{f}{h^n} = \{e^{\alpha t}\}$.

REMARK 6.1. Since $(s-[\alpha])\frac{f}{h^n} = (s-\alpha)\frac{f}{h^n} = I$ has one and only one solution $\frac{f}{h^n} = \{e^{\alpha t}\}$ in the ring $C_H$, we say that $(s-\alpha)$ admits the (*multiplicative*) *inverse* $\{e^{\alpha t}\}$ in $C_H$. We shall denote this fact by

$$(s-[\alpha])^{-1} = \frac{I}{s-[\alpha]} = \frac{1}{s-\alpha} = \{e^{\alpha t}\}. \tag{6.3}$$

In the same way, the element $\{e^{\alpha t}\} \in C \subseteq C_H$ has the (*multiplicative*) *inverse* $(s-[\alpha]) = (s-\alpha) \in C_H$:

$$\frac{I}{\{e^{\alpha t}\}} = (s-[\alpha]) = (s-\alpha). \tag{6.4}$$

COROLLARY. For any integer $n \geqq 1$, $(s-\alpha)^n = (s-[\alpha])^n \in C_H$ has the inverse in $C_H$ given by

$$\frac{1}{(s-\alpha)^n} = \frac{I}{(s-[\alpha])^n} = \left(\frac{1}{s-[\alpha]}\right)^n = \left\{\frac{t^{n-1}}{(n-1)!}\,e^{\alpha t}\right\} \quad (n = 1,2,\ldots). \tag{6.5}$$

PROOF: We have

$$\frac{1}{(s-\alpha)^2} = \frac{1}{s-\alpha}\,\frac{1}{s-\alpha} = \{e^{\alpha t}\}\{e^{\alpha t}\} = \left\{\int_0^t e^{\alpha(t-u)}e^{\alpha u}du\right\}$$
$$= \left\{e^{\alpha t}\int_0^t du\right\} = \{te^{\alpha t}\}$$

and hence

$$\frac{1}{(s-\alpha)^3} = \frac{1}{s-\alpha}\,\frac{1}{(s-\alpha)^2} = \{e^{\alpha t}\}\left\{\frac{t}{1!}e^{\alpha t}\right\} = \left\{\int_0^t e^{\alpha(t-u)}\frac{u}{1!}\,e^{\alpha u}du\right\}$$
$$= \left\{e^{\alpha t}\int_0^t \frac{u}{1!}du\right\} = \left\{\frac{t^2}{2!}\,e^{\alpha t}\right\}$$

etc.

# Chapter III
# Linear Ordinary Differential Equations with Constant Coefficients

## §7. THE CONVERSION OF THE INITIAL VALUE PROBLEM FOR THE DIFFERENTIAL EQUATION INTO A HYPERFUNCTION EQUATION

THEOREM 6. Let $\alpha_0, \alpha_1, \ldots, \alpha_n$ be complex numbers $(\alpha_n \neq 0)$ and $f \in C = C[0,\infty)$. Consider the equation

$$\alpha_n y^{(n)} + \alpha_{n-1} y^{(n-1)} + \ldots + \alpha_1 y' + \alpha_0 y = f \tag{7.1}$$

together with the *initial condition* at $t = 0$:

$$y(0) = \gamma_0, \quad y'(0) = \gamma_1, \ldots, y^{(n-1)}(0) = \gamma_{n-1}. \tag{7.2}$$

Then, (7.1)-(7.2) converts into the hyperfunction equation

$$\begin{cases} (\alpha_n s^n + \alpha_{n-1} s^{n-1} + \ldots + \alpha_1 s + \alpha_0 I) y \\ \quad = f + \beta_{n-1} s^{n-1} + \beta_{n-2} s^{n-2} + \ldots + \beta_1 s + \beta_0 I, \\ \beta_\nu = \alpha_{\nu+1}\gamma_0 + \alpha_{\nu+2}\gamma_1 + \ldots + \alpha_n \gamma_{n-\nu-1} \qquad (\nu = 0,1,2,\ldots,n-1) \end{cases} \tag{7.3}$$

by virtue of

$$y^{(k)} = s^k y - s^{k-1} y(0) - s^{k-2} y'(0) - \ldots - s y^{(k-2)}(0) - y^{(k-1)}(0), \tag{7.4}$$

where $y^{(j)}(0) = [y^{(j)}(0)]$.

REMARK 7.1. The merit of the above converstion lies in the fact that the initial condition (7.2) is absorbed in the equation (7.3). Since the terms such as $s^k y$'s are hyperfunctions derived from the unknown function $y \in C$, we call (7.3) the hyperfunction equation.

EXAMPLE 7.1. The initial value problem

$$\alpha_1 y' + \alpha_0 y = f, \quad y'(0) = \gamma_0$$

is converted into the hyperfunction equation

$$\alpha_1 sy + \alpha_0 y = f + \alpha_1\gamma_0. \tag{7.5}$$

This is equivalent to

$$\left(s + \frac{\alpha_0}{\alpha_1}\right)y = \frac{f}{\alpha_1} + \gamma_0.$$

Multiplying by $\frac{1}{s+\alpha_0/\alpha_1} = \{e^{-\alpha_0 t/\alpha_1}\}$, we obtain

$$y = \{e^{-\alpha_0 t/\alpha_1}\}\left\{\frac{f(t)}{\alpha_1} + \gamma_0\right\} = \left\{\gamma_0 e^{-\alpha_0 t/\alpha_1} + \int_0^t e^{-\alpha_0(t-u)/\alpha_1}\,\frac{f(u)}{\alpha_1}\,du\right\},$$

i.e.,

$$y(t) = \gamma_0 e^{-\alpha_0 t/\alpha_1} + \int_0^t \frac{1}{\alpha_1} e^{-\alpha_0(t-u)/\alpha_1} f(u)\,du.$$

THE SOLUTION OF THE HYPERFUNCTION EQUATION (7.3). Solving (7.3) formally we obtain

$$y = \frac{I}{\alpha_n s^n + \alpha_{n-1}s^{n-1} + \ldots + \alpha_0 s^0} f + \frac{\beta_{n-1}s^{n-1} + \beta_{n-2}s^{n-2} + \ldots + \beta_0 s^0}{\alpha_n s^n + \alpha_{n-1}s^{n-1} + \ldots + \alpha_0 s^0}. \tag{7.6}$$

In the next section, we shall give the interpretation of the right hand side of (7.6) as a function in $C[0,\infty)$.

## §8. THE POLYNOMIAL RING OF POLYNOMIALS IN s HAS NO ZERO FACTORS

Let

$$P(s) = \lambda_n s^n + \lambda_{n-1}s^{n-1} + \ldots + \lambda_0 s^0 \qquad (\lambda_n \neq 0;\ s^0 = I)$$

and

$$Q(s) = \mu_m s^n + \mu_{m-1} s^{m-1} + \ldots + \mu_0 s^0 \qquad (\mu_m \neq 0)$$

be, respectively, polynomials of n-th and m-th degree.

PROPOSITION 6. The product $P(s)Q(s)$ is a polynomial in $s$ of (n+m)-th degree.

PROOF: By the rules of the sum and product in the ring $C_H$, we obtain

$$P(s)Q(s) = \lambda_n\mu_m s^{n+m} + (\lambda_n\mu_{m-1} + \lambda_{n-1}\mu_m)s^{n+m-1}$$
$$+ \dots + \left(\sum_{i=0}^{k} \lambda_{k-i}\mu_i\right)s^k + \dots + \lambda_0\mu_0 s^0 \tag{8.1}$$

and $\lambda_n\mu_m \neq 0$.

COROLLARY. Let $P(s)$ and $Q(s)$ be polynomials in $s$. Then $P(s)Q(s) = 0$* implies that either $P(s)$ or $Q(s)$ must be $= 0$.

REMARK 8.1. The totality of polynomials in $s$ is a subring of the ring $C_H$. It is called the *polynomial ring* of $s$. The above Corollary is stated as follows:

THEOREM 7. The polynomial ring of $s$ has no *zero factors*. Precisely the product of two non-zero polynomials $P(s)$ and $Q(s)$ is not $= 0$.

DEFINITION OF THE FRACTION OF TWO POLYNOMIALS IN s (THAT IS, THE RATIONAL FUNCTIONS OF s).

PROPOSITION 7. Let, for two polynomials $p(s)$ and $q(s)$, $p(s) \neq 0$. Then we can define the "fraction" $\frac{q(s)}{p(s)}$ by the rule of the equivalence relation $\approx$:

$$\frac{q(s)}{p(s)} \approx \frac{q_1(s)}{p_1(s)} \quad \text{means} \quad q(s)p_1(s) = p(s)q_1(s), \tag{8.2}$$

because we can prove

$$\frac{q(s)}{p(s)} \approx \frac{q(s)}{p(s)},$$

$$\frac{q(s)}{p(s)} \approx \frac{q_1(s)}{p_1(s)} \quad \text{implies} \quad \frac{q_1(s)}{p_1(s)} \approx \frac{q(s)}{p(s)},$$

$$\frac{q(s)}{p(s)} \approx \frac{q_1(s)}{p_1(s)}, \quad \frac{q_1(s)}{p_1(s)} \approx \frac{q_2(s)}{p_2(s)} \quad \text{implies} \quad \frac{q(s)}{p(s)} \approx \frac{q_2(s)}{p_2(s)}.$$

PROOF: We shall give the proof of the last relation (the transitive rule). By assumption, we have

$$q(s)p_1(s) = p(s)q_1(s),$$

$$q_1(s)p_2(s) = p_1(s)q_2(s).$$

Multiplying the first formula by $p_2(s)$, we obtain

---

*$\alpha_n s^n + \dots + \alpha_0 s^0 = 0$ means that $\alpha_n = \dots = \alpha_0 = 0$. See Remark 5.3.

$$q(s)p_1(s)p_2(s) = p(s)q_1(s)p_2(s).$$

Multiplying the second formula by $p(s)$, we obtain

$$p(s)q_1(s)p_2(s) = p(s)p_1(s)q_2(s).$$

Hence

$$q(s)p_1(s)p_2(s) = p(s)p_1(s)q_2(s), \quad \text{i.e.,}$$

$$p_1(s)(q(s)p_2(s) - p(s)q_2(s)) = 0.$$

Since $p_1(s) \neq 0$ as the denominator of the "fraction", we must have

$$q(s)p_2(s) - p(s)q_2(s) = 0 \quad \text{(by Theorem 7).}$$

Therefore

$$\frac{q(s)}{p(s)} \approx \frac{q_2(s)}{p_2(s)} \; .$$

We can now replace $\approx$ by the equality sign $=$ so that

$$\frac{q(s)}{p(s)} = \frac{q_1(s)}{p_1(s)} \quad \text{means} \quad q(s)p_1(s) = p(s)q_1(s). \tag{8.2}'$$

Therefore, as in the case of Theorem 2, we can prove

THEOREM 8. The *sum* and the *product* of the "fractions" above can be defined by

$$\frac{q(s)}{p(s)} + \frac{q_1(s)}{p_1(s)} = \frac{q(s)p_1(s)+p(s)q_1(s)}{p(s)p_1(s)} \; , \quad \frac{q(s)}{p(s)} \frac{q_1(s)}{p_1(s)} = \frac{q(s)q_1(s)}{p(s)p_1(s)}. \tag{8.3}$$

Moreover, the totality of the fractions is a ring with respect to the above sum and product, and the product is commutative. We can also prove, for scalar multiplication by complex numbers,

$$\alpha\left(\frac{q(s)}{p(s)} + \frac{q_1(s)}{p_1(s)}\right) = \frac{\alpha q(s)}{p(s)} + \frac{\alpha q_1(s)}{p_1(s)} \; , \quad (\alpha+\beta)\frac{q(s)}{p(s)} = \frac{\alpha q(s)}{p(s)} + \frac{\beta q(s)}{p(s)} \; .$$

$$\alpha\left(\frac{q(s)}{p(s)} \frac{q_1(s)}{p_1(s)}\right) = \frac{\alpha q(s)q_1(s)}{p(s)p_1(s)} \; .$$

REMARK 8.2. We call the fraction $\frac{q(s)}{p(s)}$ $(p(s) \neq 0)$ a *rational function of* s. In particular, if $p(s) = s^0 = I$, we shall identify $\frac{q(s)}{s^0}$ with $q(s)$ so that a polynomial $q(s)$ may be considered as a rational function of s. Therefore, we can use the following rule:

$$q_1(s)\ \frac{q(s)}{p(s)} = \frac{q_1(s)q(s)}{s^0 p(s)} = \frac{q_1(s)q(s)}{p(s)}$$

and so

$$p(s)\ \frac{q(s)}{p(s)} = \frac{p(s)q(s)}{p(s)} = q(s).$$

## §9. THE PARTIAL FRACTION DECOMPOSITION OF A RATIONAL FUNCTION OF s

PROPOSITION 8. Given a polynomial in $s$ of n-th degree,

$$p(s) = \alpha_n s^n + \alpha_{n-1} s^{n-1} + \ldots + \alpha_0 s^0 \quad (\alpha_n \neq 0), \tag{9.1}$$

we shall consider a polynomial in the complex variable $z$ of n-th degree,

$$p(z) = \alpha_n z^n + \alpha_{n-1} z^{n-1} + \ldots + \alpha_0 z^0. \tag{9.2}$$

By a theorem of Carl Friedrich Gauss (1777-1855), $p(z)$ admits a factorization

$$p(z) = \alpha_n (z-z_1)^{m_1} (z-z_2)^{m_2} \ldots (z-z_k)^{m_k}, \quad \sum_{i=1}^{k} m_i = n, \tag{9.3}$$

where $z_1, z_2, \ldots, z_k$ are distinct roots of the algebraic equation $p(z) = 0$. Then, by the relations between the roots $z_1, z_2, \ldots, z_k$ and the coefficients $\alpha_n, \alpha_{n-1}, \ldots, \alpha_0$, we see that

$$p(s) = \alpha_n (s-z_1)^{m_1} (s-z_2)^{m_2} \ldots (s-z_k)^{m_k}. \tag{9.3)'$$

THEOREM 9. Besides $p(s)$, consider another polynomial $q(s)$ in $s$ of m-th degree with $m < n$. Then, we have the *partial fraction decomposition* of $\frac{q(s)}{p(s)}$:

$$\frac{q(s)}{p(s)} = \sum_{j=1}^{k} \sum_{u=1}^{m_j} \frac{C_{ju} I}{(s-z_j)^u} \tag{9.4}$$

with uniquely determined complex numbers $C_{ju}$'s. Therefore, we have, by (6.5),

$$\frac{q(s)}{p(s)} = \sum_{j=1}^{k} \sum_{j=1}^{m_j} C_{ji} \left\{ \frac{t^{u-1}}{(u-1)!} e^{z_j t} \right\} \tag{9.4)'$$

which is an infinitely differentiable function of $t \geqq 0$.

PROOF: By (9.3)', we have

$$p(s) = (s-z_1)^{m_1} f(s),$$

where $f(s)$ is a polynomial in $s$ of $(n-m_1)$-th degree. By (9.3), $f(z)$ does not have $(z-z_1)$ as a factor so that $f(z_1) \neq 0$ by the *remainder theorem*. Hence, the complex number

$$\mu = q(z_1)/f(z_1)$$

is well-defined. Thus, if we write

$$\frac{q(s)}{p(s)} = \frac{q(s)}{(s-z_1)^{m_1} f(s)} = \frac{\mu}{(s-z_1)^{m_1}} + \frac{q(s)-\mu f(s)}{(s-z_1)^{m_1} f(s)},$$

then

$$q(z_1) - \mu f(z_1) = 0,$$

so that, by the remainder theorem,

$$q(z) - \mu f(z) = (z-z_1) f_1(z).$$

Since both $q(z)$ and $f(z)$ are of degree $< n$, the degrees in $s$ of $(s-z_1)^{m_1} f(s)$ and $f_1(s)$ are respectively $n$ and $< n-1$. Hence, if $m_1 - 1 > 0$, we have

$$\frac{q(s)-\mu f(s)}{(s-z_1)^{m_1} f(s)} = \frac{(s-z_1) f_1(s)}{(s-z_1)^{m_1} f(s)} = \frac{f_1(s)}{(s-z_1)^{m_1-1} f(s)},$$

where the degrees in $s$ of $(s-z_1)^{m_1-1} f(s)$ and $f_1(s)$ are respectively $(n-1)$ and $< (n-1)$.

Therefore, we are able to start with

$$\frac{f_1(s)}{(s-z_1)^{m_1-1} f(s)}$$

and argue as above again. Repeating the same process, we can finally obtain (9.4).

REMARK 9.1. The uniqueness of the $C_{jk}$'s of (9.4) is proved as follows. Multiplying $(z-z_1)^{m_1}$ by

$$\frac{q(z)}{p(z)} = \sum_{j=1}^{k} \sum_{u=1}^{m_j} \frac{C_{ju}}{(z-z_j)^u},$$

we obtain

$$\lim_{z \to z_1} (z-z_1)^{m_1} \frac{q(z)}{p(z)} = C_{1,m_1}. \tag{9.5}$$

Next we obtain

$$\lim_{z \to z_1} (z-z_1)^{m_1-1}\left(\frac{q(z)}{p(z)} - \frac{C_{1,m_1}}{(z-z_1)^{m_1}}\right) = C_{1,m_1-1}, \tag{9.6}$$

$$\lim_{z \to z_1} (z-z_1)^{m_1-2}\left(\frac{q(z)}{p(z)} - \frac{C_{1,m_1}}{(z-z_1)^{m_1}} - \frac{C_{1,m_1-1}}{(z-z_1)^{m_1-1}}\right) = C_{1,m_1-2}, \tag{9.7}$$

and so forth.

EXAMPLE 9.1. Obtain a, b, and c in

$$\frac{s+\beta}{s^2(s+\alpha)} = \frac{a}{s^2} + \frac{b}{s} + \frac{c}{s+\alpha}.$$

ANSWER:

$$a = \lim_{z\to 0} \frac{z^2(z+\beta)}{z^2(z+\alpha)} = \frac{\beta}{\alpha},$$

$$b = \lim_{z\to 0} z\left(\frac{z+\beta}{z^2(z+\alpha)} - \frac{\beta}{\alpha z^2}\right) = \lim_{z\to 0} \frac{\alpha z+\alpha\beta-\beta(z+\alpha)}{\alpha z(z+\alpha)} = \frac{\alpha-\beta}{\alpha^2},$$

$$c = \lim_{z\to -\alpha} \frac{(z+\alpha)(z+\beta)}{z^2(z+\alpha)} = \frac{\beta-\alpha}{\alpha^2}.$$

EXAMPLE 9.2. Obtain the partial fraction decomposition of

$$\frac{s^4}{s^3-1} = s + \frac{s}{s^3-1}.$$

ANSWER: We have

$$\frac{z}{z^3-1} = \frac{a}{z-1} + \frac{b}{z-\omega} + \frac{c}{z-\omega^2}, \quad \omega = \frac{-1+i\sqrt{3}}{2}, \quad \omega^2 = \frac{-1-i\sqrt{3}}{2},$$

and so

$$a = \lim_{z\to 1} \frac{(z-1)z}{(z^3-1)} = \lim_{z\to 1} \frac{z}{z^2+z+1} = \frac{1}{3},$$

$$b = \lim_{z\to\omega} \frac{(z-\omega)z}{(z^3-\omega^3)} = \lim_{z\to\omega} \frac{z}{z^2+\omega z+\omega^2} = \frac{\omega^2}{3},$$

$$c = \lim_{z\to\omega^2} \frac{(z-\omega^2)z}{(z^3-(\omega^2)^3)} = \lim_{z\to\omega^2} \frac{z}{z^2+\omega^2 z+\omega^4} = \frac{\omega}{3},$$

$$\frac{s}{s^3-1} = \frac{1}{3}\left\{e^t + \omega^2 e^{\omega t} + \omega e^{\omega^2 t}\right\}.$$

EXAMPLE 9.3. Let $\alpha$ and $\beta$ be real numbers. Prove (i) and (ii) below.

(i) $$\frac{I}{(s-\alpha)^2+\beta^2} = \{\frac{1}{\beta} e^{\alpha t} \sin \beta t\} \qquad (\beta \neq 0).$$

(ii) $$\frac{s-\alpha}{(s-\alpha)^2+\beta^2} = \{e^{\alpha t} \sin \beta t\}.$$

PROOF:

$$\frac{I}{(s-\alpha)^2+\beta^2} = \frac{1}{2i\beta}\left(\frac{I}{s-\alpha-i\beta} - \frac{I}{s-\alpha+i\beta}\right)$$

$$= \frac{1}{2i\beta}\{e^{(\alpha+i\beta)t} - e^{(\alpha-i\beta)t}\} = \{\frac{1}{\beta} e^{\alpha t} \sin\beta t\},$$

$$\frac{s-\alpha}{(s-\alpha)^2+\beta^2} = \frac{1}{2}\left(\frac{I}{s-\alpha-i\beta} + \frac{I}{s-\alpha+i\beta}\right)$$

$$= \frac{1}{2}\{e^{(\alpha+i\beta)t} + e^{(\alpha-t\beta)t}\} = \{e^{\alpha t} \cos \beta t\}.$$

EXAMPLE 9.4. By (i) of Example 9.3, we have

$$\frac{I}{[(s-\alpha)^2+\beta^2]^2} = \left\{\frac{1}{\beta^2}\int_0^t e^{\alpha(t-\tau)}\sin \beta(t-\tau)e^{\alpha\tau} \sin \beta\tau d\tau\right\}$$

$$= \left\{\frac{-e^{\alpha t}}{2\beta^2}\int_0^t (\cos \beta t - \cos \beta(t-2\tau))d\tau\right\}$$

$$= \left\{\frac{e^{\alpha t}}{2\beta^2}\left(\frac{1}{\beta} \sin \beta t - t \cos \beta t\right)\right\}.$$

As a remark to Theorem 9, we have

THEOREM 9'. Let $p(s)$ and $q(s)$ be polynomials in $s$ of degree $n$ and $m$ respectively, with $n \leq m$. Then

$$\frac{q(s)}{p(s)} = w(s) + \frac{r(s)}{p(s)}, \tag{9.8}$$

where $w(s)$ is a polynomial in $s$ of degree $m-n$ and $r(s)$ is a polynomial in $s$ of degree $\leq n-1$.

PROOF: We omit the proof.

COROLLARY. The polynomial $w(s)$ is an element of $C_H$ and the fraction $\frac{r(s)}{p(s)}$ is, by (9.4)', an infinitely differentiable function in $C$.

EXAMPLE 9.5.

$$\frac{s^3}{s^2-1} = s + \frac{s}{s^2-1} = s + \frac{1}{2}\left(\frac{I}{s-1} + \frac{I}{s+1}\right)$$
$$= s + \left\{\frac{e^t + e^{-t}}{2}\right\}.$$

EXERCISES FOR §9. Prove the following formulas ($\alpha$ and $\beta$ are real numbers).

1. $\dfrac{I}{(s-1)^2(s+1)(s-2)} = \dfrac{1}{12}\{-6te^t - 3e^t - e^{-t} + 4e^{2t}\},$

2. $\dfrac{I}{(s-\alpha)^2-\beta^2} = \{\dfrac{1}{\beta} e^{\alpha t} \sinh(\beta t)\} \quad (\beta \neq 0),$

3. $\dfrac{s-\alpha}{(s-\alpha)^2-\beta^2} = \{e^{\alpha t} \cosh(\beta t)\}.$*

## §10. HYPERFUNCTION SOLUTION OF THE ORDINARY DIFFERENTIAL EQUATION (THE OPERATIONAL CALCULUS)

The initial value problem of §7:

$$\alpha_n y^{(n)} + \alpha_{n-1} y^{(n-1)} + \ldots + \alpha_1 y' + \alpha_0 y = f, \tag{10.1}$$

$$y(0) = \gamma_0, \quad y'(0) = \gamma_1, \ldots, y^{(n-1)}(0) = \gamma_{n-1}, \tag{10.2}$$

was converted into the hyperfunction equation

$$\begin{cases} (\alpha_n s^n + \alpha_{n-1} s^{n-1} + \ldots + \alpha_1 s + \alpha_0 I) y \\ \qquad = f + \beta_{n-1} s^{n-1} + \beta_{n-2} s^{n-2} + \ldots + \beta_1 s + \beta_0 I \\ \beta_\nu = \alpha_{\nu+1}\gamma_0 + \alpha_{\nu+2}\gamma_1 + \ldots + \alpha_n \gamma_{n-\nu-1} \quad (\nu = 0,1,2,\ldots,n-1). \end{cases} \tag{10.3}$$

Let us denote

$$p(s) = \alpha_n s^n + \alpha_{n-1} s^{n-1} + \ldots + \alpha_0 s^0 \quad (s^0 = I). \tag{10.4}$$

Then we have the factorization

$$p(z) = \alpha_n (z-z_1)^{m_1} (z-z_2)^{m_2} \ldots (z-z_k)^{m_k}, \quad \sum_{j=1}^{k} m_j = n, \tag{10.5}$$

---

*$\sinh t = \sinh(t) = (e^t - e^{-t})/2$, $\cosh t = \cosh(t) = (e^t + e^{-t})/2$.

and so, by (9.4)',

$$\frac{I}{p(s)} = \sum_{j=1}^{k}\sum_{u=1}^{m_j} C_{ju}\left\{\frac{t^{u-1}}{(u-1)!}e^{z_j t}\right\}, \tag{10.6}$$

and

$$\begin{cases} \dfrac{r(s)}{p(s)} = \displaystyle\sum_{j=1}^{k}\sum_{u=1}^{m_j} d_{ju}\left\{\frac{t^{u-1}}{(u-1)!}e^{z_j t}\right\}, \\ r(s) = \beta_{n-1}s^{n-1} + \beta_{n-2}s^{n-2} + \dots + \beta_1 s + \beta_0 s^0. \end{cases} \tag{10.7}$$

Therefore, by multiplying both sides of (10.3) by $\frac{I}{p(s)}$, we obtain

$$\begin{aligned} y &= \frac{I}{p(s)} f + \frac{r(s)}{p(s)} \\ &= \sum_{j=1}^{k}\sum_{u=1}^{m_j}\left[C_{ju}\left\{\frac{t^{u-1}}{(u-1)!}e^{z_j t}\right\}f + d_{ju}\left\{\frac{t^{u-1}}{(u-1)!}e^{z_j t}\right\}\right]. \end{aligned} \tag{10.8}$$

REMARK 10.1. If $f \in C[0,\infty)$, then $y = y(t)$ given by (10.8) is in $C[0,\infty)$. In fact, we can prove that this $y(t)$ is *n-times continuously differentiable in* $t$. Therefore, this $y(t)$ *is the unique solution of (10.1) - (10.2).*

PROOF OF THE n-TIMES CONTINUOUS DIFFERENTIABILITY OF (10.8): Multiplying both sides of (10.3) by $h^n$, we obtain

$$\alpha_n y + \alpha_{n-1}hy + \dots + \alpha_0 h^n y = \{h^n f\} + \beta_{n-1}h + \dots + \beta_0 h^n.$$

Here

$$F(t) = h^n f + \beta_{n-1}h + \dots + \beta_0 h^n$$

is n-times continuously differentiable in $t$, and so, $y$ given by

$$y = -\alpha_n^{-1}(\alpha_{n-1}hy + \alpha_{n-2}h^2 y + \dots + \alpha_0 h^n y) + \alpha_n^{-1}\{F(t)\} \tag{10.9}$$

is once continuously differentiable in $t$ (because, for example, $(h^3 y)$ is, by the continuity of $y(t)$, once continuously differentiable; and furthermore, by (5.3)',

$$\begin{aligned} (h^3 y)' &= h^2 y = h^2(hy' + y(0)) = h^3 y' + h^2 y(0) \\ &= h^3 y' + (\text{a polynomial in } t).) \end{aligned}$$

Therefore, by differentiating (10.9) once, we obtain

$$y' = -\alpha_n^{-1}(\alpha_{n-1}hy' + \alpha_{n-2}h^2y' + \dots + \alpha_0 h^n y') \qquad (10.9)'$$
$$+ \text{(a polynomial in } t) + \alpha_n^{-1}F'(t).$$

Repeating the above reasoning on (10.9)', we can prove that $y$, given by (10.8), is n-times continuously differentiable in $t$ whenever $f = \{f(t)\} \in C$.

Thus we are now able to state:

THE OPERATIONAL METHOD OF SOLVING (10.1)-(10.2). Remembering the formula

$$y' = sy - [y(0)],$$

we easily extend it to

$$y'' = sy' - [y'(0)] = s(sy - [y(0)]) - [y'(0)] = s^2y - sy(0) - y'(0),$$

and finally to

$$y^{(k)} = s^ky - s^{k-1}y(0) - s^{k-2}y'(0) - \dots - y^{(k-1)}(0).$$

By virtue of this formula, we obtain (10.3) from the initial value problem (10.1) - (10.2). Hence we have $p(s)$ ((10.4)) and $r(s)$ ((10.7)). Therefore, we obtain the solution $y$ of (10.1)-(10.2):

$$y = \frac{I}{p(s)} f + \frac{r(s)}{p(s)} .$$

Then, by making use of the partial fraction decomposition of

$$\frac{I}{p(s)} \quad \text{and} \quad \frac{q(s)}{p(s)}$$

together with the formula

$$\frac{I}{(s-z_j)^u} = \left\{\frac{t^{u-1}}{(u-1)!} e^{z_jt}\right\},$$

we obtain a concrete expression of the solution $y(t)$ of (10.1)-(10.2).

EXAMPLE 10.1. Solve $y' + \alpha y = f$, $y(0) = a$. By (7.4), we have

$$sy - a + \alpha y = f$$

so that

$$y = \frac{I}{s+\alpha} f + \frac{a}{s+\alpha} = \{e^{-\alpha t}\}f + \{ae^{-\alpha t}\}$$
$$= \left\{e^{-\alpha t} \int_0^t e^{\alpha u}f(u)du + ae^{-\alpha t}\right\}.$$

REMARK 10.2. The solution of $y' - y = \{(2t-1)e^{t^2}\}$, $y(0) = 2$ is given by

$$y = \left\{e^t \int_0^t e^{-u}(2u-1)e^{u^2} du + 2e^t\right\} = \left\{e^t + e^{t^2}\right\},$$

because

$$\int_0^t e^{-u}(2u-1)e^{u^2} du = e^{-u}e^{u^2}\Big|_{u=0}^{u=t} + \int_0^t e^{-u}e^{u^2} du - \int_0^t e^{-u}e^{u^2} du$$
$$= e^{-t}e^{t^2} - 1.$$

It is to be noted that the above equation is difficult to solve by the customary method of the *Laplace transform,* because

$$\int_0^\infty e^{\lambda t}(2t-1)e^{t^2} dt$$

diverges for any $\lambda$.

EXAMPLE 10.2. Solve $y' + \alpha y = \beta$, $y(0) = a$. Since $\beta$ means

$$\{\beta\} = [\beta]h = \frac{[\beta]}{s} = \frac{\beta}{s}, \;{}^*$$

the corresponding hyperfunction equation is

$$sy - a + \alpha y = \frac{[\beta]}{s} = \frac{\beta}{s}.$$

Hence

$$y = \frac{1}{s+\alpha}\frac{\beta}{s} + \frac{a}{s+\alpha} = \{e^{-\alpha t}\}\{\beta e^0\} + \{ae^{-\alpha t}\}$$
$$= \left\{e^{-\alpha t}\int_0^t \beta e^{\alpha u} du + ae^{-\alpha t}\right\} = \left\{\frac{\beta}{\alpha} + (a - \frac{\beta}{\alpha})e^{-\alpha t}\right\}.$$

EXAMPLE 10.3. Solve $y'' - 4y = e^{2t}$, $y(0) = 1$, $y'(0) = \frac{1}{4}$. The corresponding hyperfunction equation is

$$s^2y - s - \frac{1}{4} - 4y = \frac{1}{s-2},$$

that is,

$$(s^2-4)y = \frac{1}{s-2} + \frac{1}{4} + s = \frac{1}{4}\frac{s+2}{s-2} + s.$$

Hence, by (10.9),

---

* As was stated in §5, we shall write $\beta/s$ for $[\beta]/s$ in case confusion will not occur. We also write 1 for I.

$$y = \frac{1}{4}\,\frac{1}{s^2-4}\,\frac{s+2}{s-2} + \frac{s}{s^2-4} = \frac{1}{4}\,\frac{1}{(s-2)^2} + \frac{s}{s^2-4}$$

$$= \frac{1}{4}\{te^{2t}\} + \frac{1}{2}\left(\frac{1}{s-2} + \frac{1}{s+2}\right) = \{\frac{1}{4}\,te^{2t} + \cosh 2t\}.$$

EXAMPLE 10.4. Let $\alpha,\beta$ be real numbers. Solve the system of linear ordinary differential equations

$$\begin{cases} y' - \alpha y - \beta z = \beta e^{\alpha t}, & z' + \beta y - \alpha z = 0, \\ y(0) = 0, & z(0) = 1. \end{cases}$$

The corresponding system of hyperfunction equations is

$$sy - \alpha y - \beta z = \frac{\beta}{s-\alpha}, \quad sz - 1 + \beta y - \alpha z = 0.$$

We solve this system, obtaining

$$y = \frac{2\beta}{(2-\alpha)^2+\beta^2}, \quad z = \frac{(s-\alpha)^2-\beta^2}{(s-\alpha)((s-\alpha)^2+\beta^2)}.$$

Hence, by (i)-(ii) of Example 9.3, we have

$$y = \{2e^{\alpha t}\sin\beta t\}$$

$$z = \frac{2(s-\alpha)}{(s-\alpha)^2+\beta^2} - \frac{1}{s-\alpha} = \{2e^{\alpha t}\cos\beta t - e^{\alpha t}\}.$$

EXAMPLE 10.5. Solve $y' - y = 2$, $y(1) = 1$. Putting $y(t+1) = z(t)$, we obtain

$$z' - z = 2, \quad z(0) = 1.$$

The corresponding hyperfunction equation is

$$sz - 1 - z = \frac{2}{s},$$

so that, by (6.5), we obtain

$$z = \frac{1}{s-1} + \frac{2}{(s-1)s} = \{e^t\} + 2\left(\frac{1}{s-1} - \frac{1}{s}\right)$$

$$= \{e^t + 2e^t - 2\} = \{3e^t - 2\}.$$

Hence $y(t) = z(t-1) = \{3e^{t-1} - 2\}$.

EXAMPLE 10.6. Solve $y'' + 2ay' + b^2y = f(t)$, $y(0) = k_0$, $y'(0) = k_1$. Here $a,b,k_0,k_1$ are real constants with $a > 0$ and $b > 0$. The corresponding hyperfunction equation is

$$s^2y + 2asy + b^2y = \{f(t)\} + sk_0 + k_1 + 2ak_0,$$

and so

$$y = \frac{f}{s^2+2as+b^2} + \frac{sk_0+k_1+2ak_0}{s^2+2as+b^2}.$$

For the sake of convenience, we introduce

$$\omega^2 = b^2 - a^2.$$

Then, if $\omega \neq 0$, we have

$$y = \frac{f}{(s+a+i\omega)(s+a-i\omega)} + \frac{(s+a)k_0+(ak_0+k_1)}{(s+a+i\omega)(s+a-i\omega)}$$

$$= \frac{-1}{2i\omega}\left(\frac{1}{s+a+i\omega} - \frac{1}{s+a-i\omega}\right)(f+ak_0+k_1)$$

$$+ \frac{k_0}{2}\left(\frac{1}{s+a+i\omega} + \frac{1}{s+a-i\omega}\right)$$

$$= e^{-at}\frac{e^{i\omega t}-e^{-i\omega t}}{2i\omega}(f+ak_0+k_1) + k_0e^{-at}\frac{e^{i\omega t}+e^{-i\omega t}}{2}$$

$$= \left\{\frac{e^{-at}}{\omega}\int_0^t e^{a\tau}\sin\omega(t-\tau)f(\tau)d\tau + \frac{ak_0+k_1}{\omega}e^{-at}\sin\omega t + k_0e^{-at}\cos\omega t\right\}.$$

REMARK 10.3. In the case $\omega = 0$, i.e., $b^2 = a^2$, we obtain

$$y = \frac{f+ak_0+k_1}{(s+a)^2} + \frac{(s+a)k_0}{(s+a)^2}$$

$$= \{te^{-at}\}\{f(t)\} + \{(ak_0+k_1)te^{-at} + k_0e^{-at}\}.$$

The equation of Example 10.6 represents the motion of a *particle of unit mass*. The left side of the equation is the sum of the *acceleration*, the *damping force* and the *elastic force*, and the right side of the equation is the *external force* acting on the particle.

EXERCISES FOR §10. Solve the given differential equations (systems).

1. $y'' + 3y' + 2y = 2te^{-t}$, $y(0) = y'(0) = 0$.
2. $y'' + 3y' + 2y = 2te^{-t}$, $y(1) = y'(1) = 0$ (Hint: Put $y(t+1) = z(t)$).
3. $y''' + y' = e^{2t}$, $y(0) = y'(0) = y''(0) = 0$.
4. $y' + z' - z = e^t$, $2y' + z' + 2z = \cos t$; $x(0) = y(0) = 0$.
5. $x' = y-z$, $y' = x+y$, $z' = x+z$; $x(0) = 1$, $y(0) = 2$, $z(0) = 3$.

6. $x'-x-2y = t, \quad -2x+y'-y = t, \quad x(0) = 2, \quad y(0) = 4.$

7. $x'' + 3y' - 4x + 6y = 10 \cos t,$

$x' + y'' - 2x + 4y = 0,$

$x(0) = y(0) = 0, \quad x'(0) = 4, \quad y'(0) = 2.$

EXAMPLE 10.7. CONCRETE EXAMPLES (ELECTRIC CIRCUITS). A circuit of self-inductance $L$ and resistance $R$ has applied an electromotive force $E$. If $i(t)$ denotes the current, we have, by Kirchhoff's law,

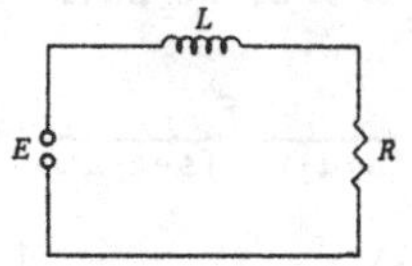


$$Li'(t) + Ri(t) = E,$$

where $E$ is assumed to be $E = \{E_0\}$. Assuming the initial condition of the current $i(t)$ to be $i(0) = 0$, we obtain

$$Lsi + Ri = \frac{E_0}{s}.$$

Hence

$$s(Ls+R)i = E_0$$

and so

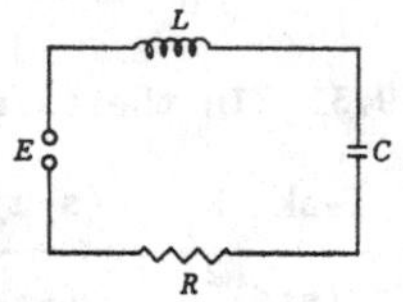


$$i = E_0 \frac{1}{s(Ls+R)} = \frac{E_0}{R}\left(\frac{1}{s} - \frac{1}{s+R/L}\right),$$

that is,

$$i(t) = \frac{E_0}{R}\{1 - e^{-(R/L)t}\}.$$

EXAMPLE 10.8. Consider a circuit of self-inductance $L$, resistance $R$ and of capacitance $C$. Let the charge at the electric pole of the condenser be $Q$. Then

$$Li'(t) + Ri(t) + \frac{Q(t)}{C} = E,$$

$$Q'(t) = i(t).$$

Assuming the initial conditions $i(0) = 0$ and $Q(0) = 0$, we have

$$Lsi + Ri + \frac{Q}{C} = E, \quad sQ = i.$$

Eliminating $Q$ from these two equations, we obtain, taking $E = \{E_0\}$,

$$\left(Ls + R + \frac{I}{Cs}\right)i = E = \frac{[E_0]}{s}.$$

Thus

$$\begin{cases} i = \dfrac{[E_0]}{L(s^2 + \frac{R}{L}s + \frac{I}{LC})} = \dfrac{[E_0]}{L((s+\alpha)^2 + \mu)}, \\ \alpha = \dfrac{R}{2L}, \quad \mu = \dfrac{1}{LC} - \dfrac{R^2}{4L^2}. \end{cases}$$

Hence, by (6.5), (i) of Example 9.3, and 2 of Exercises in §9, we obtain

$$i(t) = \begin{cases} \dfrac{E_0}{L}\dfrac{1}{\sqrt{\mu}}e^{-\alpha t}\sin\sqrt{\mu}t & (\mu > 0) \\ \dfrac{E_0}{L}te^{-\alpha t} & (\mu = 0) \\ \dfrac{E_0}{L}\dfrac{1}{\sqrt{-\mu}}e^{-\alpha t}\sinh\sqrt{-\mu}t & (\mu < 0). \end{cases}$$

## §11. BOUNDARY VALUE PROBLEMS FOR ORDINARY DIFFERENTIAL EQUATIONS

PROBLEM 11.1. Let $\alpha \neq 0$ be a real number, $t_0 > 0$ and $f(t) \in C[0,t_0]$, the totality of complex-valued continuous functions defined on the closed interval $[0,t_0]$. Find the solution of the boundary value problem:

$$\begin{cases} y''(t) + \alpha^2 y(t) = f(t) & (0 \leqq t \leqq t_0), \\ y(0) = y(t_0) = 0, \end{cases} \tag{11.1}$$

ANSWER: Define the values of $f$ for $t \geqq t_0$ by $f(t) = f(t_0)$. Then $f \in C = C[0,\infty)$. We first obtain the solution of

$$y''(t) + \alpha^2 y(t) = f(t), \quad y(0) = 0, \quad y'(0) = \beta. \tag{11.2}$$

The hyperfunction equation for (11.2) is

$$s^2 y - \beta + \alpha^2 y = f,$$

and its solution $y$ is given by

$$\begin{aligned} y &= \frac{\beta}{s^2+\alpha^2} + \frac{f}{s^2+\alpha^2} \\ &= \frac{\beta}{2i\alpha}\left(\frac{I}{s-i\alpha} - \frac{I}{s+i\alpha}\right) + \frac{1}{2i\alpha}\left(\frac{I}{s-i\alpha} - \frac{I}{s+i\alpha}\right)f \\ &= \frac{\beta}{\alpha}\left\{\frac{e^{i\alpha t}-e^{-i\alpha t}}{2i}\right\} + \frac{1}{\alpha}\left\{\frac{e^{i\alpha t}-e^{-i\alpha t}}{2i}\right\}f \end{aligned}$$

$$= \frac{\beta}{\alpha}\{\sin \alpha t\} + \frac{1}{\alpha}\left\{\int_0^t \sin \alpha(t-u)f(u)du\right\},$$

that is,

$$y(t) = \frac{\beta}{\alpha}\{\sin \alpha t\} + \frac{1}{\alpha}\left\{\int_0^t \sin \alpha(t-u)f(u)du\right\}. \tag{11.3}$$

This $y(t)$ is a solution of (11.1) if $y(t_0) = 0$. To discuss this condition, we distinguish two cases:

i) the first case: $\sin \alpha t_0 \neq 0$. In this case, $y(t_0) = 0$ is equivalent to

$$\frac{\beta}{\alpha} \sin \alpha t_0 + \frac{1}{\alpha}\int_0^{t_0} \sin \alpha(t_0-u)f(u) = 0. \tag{11.4}$$

$\beta$ is uniquely determined since $\sin \alpha t_0 \neq 0$. Thus, by taking this $\beta$ in (11.3), we obtain the solution of (11.1).

ii) the second case: $\sin \alpha t_0 = 0$. In this case, we must have

$$\int_0^{t_0} \sin \alpha(t_0-u)f(u) = 0$$

from $y(t_0) = 0$. Thus, by

$$\sin \alpha(t_0-u) = \sin \alpha t_0 \cdot \cos \alpha u - \cos \alpha t_0 \cdot \sin \alpha u = -\cos \alpha t_0 \cdot \sin \alpha u,$$

we must have

$$\int_0^{t_0} \sin \alpha u \cdot f(u)du = 0. \tag{11.5}$$

And if (11.5) holds (of course, under the condition $\sin \alpha t_0 = 0$), $y(t)$ given by (11.3) is a solution of (11.1) for all $\beta$.

Therefore, under the condition $\sin \alpha t_0 = 0$, if $f$ does not satisfy (11.5) then (11.1) has no solution.

REMARK 11.1. When $f(t) \equiv 0$, (11.5) holds and so for those $\alpha$ for which $\sin \alpha t_0 = 0$, (11.1) has a solution.

Hence, those $\alpha \neq 0$ with $\sin \alpha t_0 = 0$ are characterized by the property that the equation

$$y''(t) + \alpha^2 y(t) = 0; \quad y(0) = 0, \quad y(t_0) = 0 \tag{11.6}$$

admits a *non-zero solution* $y(t)$, i.e., the real number $\alpha^2 \neq 0$ is an *eigenvalue* of the boundary value problem (11.6) and the non-zero solution $y(t)$ is the corresponding *eigenfunction* of (11.6).

PROBLEM 11.2. Is the following problem solvable?

$$y''(t) + \alpha^2 y(t) = 0, \quad y(0) = 0, \quad y(2\pi) = 1. \tag{11.7}$$

ANSWER: Any $y$ satisfying (11.7) is a solution of (11.2) with $f(t) \equiv 0$. Such a function is $y = \alpha^{-1}\beta \sin \alpha t$. Hence, if (11.7) has a solution, then $2\alpha$ must not be an integer. And when $2\alpha$ is not an integer, the solution of (11.7) is given by

$$y(t) = \frac{\sin \alpha t}{\sin 2\alpha\pi}.$$

PROBLEM 11.3. Is the following problem solvable?

$$y''(t) - \alpha^2 y(t) = 0, \quad y(0) = 0, \quad y(2\pi) = 1. \tag{11.8}$$

ANSWER: We first solve the initial value problem

$$y''(t) - \alpha^2 y(t) = 0, \quad y(0) = 0, \quad y'(0) = \beta,$$

obtaining

$$y = \frac{\beta}{s^2-\alpha^2} = \frac{\beta}{2a}\left(\frac{I}{s-\alpha} - \frac{I}{s+\alpha}\right) = \frac{\beta}{\alpha}\left\{\frac{e^{\alpha t} - e^{-\alpha t}}{2}\right\}.$$

This $y$ satisfies $y(2\pi) = 1$ if

$$\beta = \frac{\alpha}{\sinh(2\alpha\pi)}, \quad y(t) = \frac{\sinh(\alpha t)}{\sinh(2\alpha\pi)}.$$

EXERCISES FOR §11. Determine the functions $y$ which satisfy the given differential equations and boundary conditions.

1. $y'' + \alpha^2 y = 0;$ $y(0) = 0,$ $y'(2\pi) = 1.$
2. $y'' - \alpha^2 y = 0;$ $y(0) = 0,$ $y'(2\pi) = 1.$

($\alpha$ is a real number).

# Chapter IV
# Fractional Powers of Hyperfunctions $h$, $s$ and $\dfrac{I}{s-\alpha}$

## §12. EULER'S INTEGRALS - THE GAMMA FUNCTION AND BETA FUNCTION

These functions are respectively defined by Euler's* integrals:

$$\Gamma(\lambda) = \int_0^\infty t^{\lambda-1}e^{-t}dt \qquad (\mathrm{Re}\,\lambda > 0)^{**}, \tag{12.1}$$

$$B(\lambda,\mu) = \int_0^1 t^{\lambda-1}(1-t)^{\mu-1}dt \qquad (\mathrm{Re}\,\lambda > 0,\ \mathrm{Re}\,\mu > 0). \tag{12.2}$$

THEOREM 10.

$$\Gamma(\lambda+1) = \lambda\Gamma(\lambda), \tag{12.3}$$

$$\Gamma(n) = (n-1)! \qquad (n = 1,2,\ldots), \tag{12.4}$$

$$B(\lambda,\mu) = \frac{\Gamma(\lambda)\Gamma(\mu)}{\Gamma(\lambda+\mu)}, \tag{12.5}$$

$$\Gamma(\tfrac{1}{2}) = \sqrt{\pi}. \tag{12.6}$$

PROOF: (12.3):

$$\Gamma(\lambda) = \int_0^\infty t^{\lambda-1}e^{-t}dt = \left[\frac{t^\lambda}{\lambda}e^{-t}\right]_{t=0}^\infty + \frac{1}{\lambda}\int_0^\infty t^\lambda e^{-t}dt = \frac{1}{\lambda}\Gamma(\lambda+1).$$

(12.4): Combine (12.3) with

$$\Gamma(1) = \int_0^\infty e^{-t}dt = [-e^{-t}]_{t=0}^\infty = 1.$$

---

* Leonhard Euler (1707-1783).

** $t^{\lambda-1} = \exp((\lambda-1)\log t)$ taking the *principal value* of $\log t$ for $t > 0$ so that $|t^{\lambda-1}| = \exp((\mathrm{Re}\,\lambda-1)\log t)$. Hence the integrals in (12.1) and in (12.2) are convergent.

(12.5):

$$\Gamma(\lambda)\Gamma(\mu) = \int_0^\infty\int_0^\infty t^{\lambda-1}e^{-t}u^{\mu-1}e^{-\mu}dtdu.$$

By the substitution

$$t = \tau\nu, \quad u = (1-\tau)\nu \qquad (0 < \tau < 1,\ \nu > 0) \tag{12.7}$$

with the Jacobian determinant

$$\frac{D(t,u)}{D(\tau,\nu)} = \begin{vmatrix} \frac{\partial\tau\nu}{\partial\tau}, & \frac{\partial(1-\tau)\nu}{\partial\tau} \\ \frac{\partial\tau\nu}{\partial\nu}, & \frac{\partial(1-\tau)\nu}{\partial\nu} \end{vmatrix} = \begin{vmatrix} \nu, & -\nu \\ \tau, & (1-\tau) \end{vmatrix} = \nu,$$

we obtain

$$\Gamma(\lambda)\Gamma(\mu) = \int_0^1 d\tau\int_0^\infty d\nu\{(\tau\nu)^{\lambda-1}e^{-\tau\nu}((1-\tau)\nu)^{\mu-1}e^{-(1-\tau)\nu}\ \nu\}$$

$$= \int_0^1 \tau^{\lambda-1}(1-\tau)^{\mu-1}d\tau\int_0^\infty \nu^{\lambda+\mu-1}e^{-\nu}d\nu = B(\lambda,\mu)\Gamma(\lambda+\mu).$$

Hence we have (12.5).*

(12.6): By (12.4) and (12.5),

$$\Gamma(\tfrac{1}{2})\Gamma(\tfrac{1}{2}) = \Gamma(\tfrac{1}{2} + \tfrac{1}{2}) \int_0^1 t^{-1/2}(1-t)^{-1/2}dt$$

$$= 1\cdot\int_0^1 \frac{1}{\sqrt{t(1-t)}}\,dt = \int_0^1 \frac{2xdx}{\sqrt{x^2(1-x^2)}}$$

$$= \int_0^1 \frac{2dx}{\sqrt{1-x^2}} = 2[\arcsin x]_{x=0}^{x=1} = \pi$$

so that $\Gamma(\frac{1}{2}) = \pm\sqrt{\pi}$. We take the plus sign since $\Gamma(\lambda)$ $(\lambda > 0)$ is positive, being an integral of a positive-valued continuous function.

COROLLARY OF (12.6).

$$\int_0^\infty e^{-t^2}dt = \frac{\sqrt{\pi}}{2}\,. \tag{12.8}$$

---

*We omit the proof that $\Gamma(\lambda) \neq 0$ whenever $\operatorname{Re}\lambda > 0$. More precisely, it is proved in analytic function theory that $1/\Gamma(\lambda)$ $(\operatorname{Re}\lambda > 0)$ can be analytically continued to an *entire function*, i.e., to a regular analytic function defined for every complex number $\lambda$. See, e.g., E. Hille: Analytic Function Theory, Vol. I, Ginn and Company (1959), 229-.

PROOF: By the substitution $t = \tau^2$, we have

$$\Gamma(\tfrac{1}{2}) = \int_0^\infty \tau^{-1} e^{-\tau^2} \cdot 2\tau d\tau = 2\int_0^\infty e^{-\tau^2} d\tau.$$

COROLLARY OF (12.5). If $n$ is a positive integer,

$$B(\lambda,n) = \frac{1\cdot 2\cdot \ldots \cdot (n-1)}{\lambda(\lambda+1)\ldots(\lambda+n-1)} \qquad (\text{Re } \lambda > 0). \tag{12.9}$$

PROOF:

$$B(\lambda,1) = \int_0^1 t^{\lambda-1}(1-t)^0 dt = \frac{1}{\lambda}.$$

Then, integrating by parts,

$$B(\lambda,n) = \int_0^1 t^{\lambda-1}(1-t)^{n-1}dt = \left[\frac{t^\lambda}{\lambda}(1-t)^{n-1}\right]_{t=0}^{t=1} + \int_0^1 \frac{t^\lambda}{\lambda}(n-1)(1-t)^{n-2}dt$$

$$= \frac{n-1}{\lambda}\int_0^1 t^\lambda(1-t)^{n-2}dt = \frac{n-1}{\lambda} B(\lambda+1,n-1).$$

Therefore,

$$B(\lambda,n) = \frac{n-1}{\lambda}\frac{n-2}{\lambda+1} B(\lambda+2,n-2)$$

$$= \frac{n-1}{\lambda}\frac{n-2}{\lambda+1}\frac{n-3}{\lambda+2}\cdots\frac{1}{\lambda+n-2} B(\lambda+n-1,1)$$

$$= \frac{(n-1)(n-2)\cdots 1}{\lambda(\lambda+1)\cdots(\lambda+n-1)} \qquad (B(\lambda,1) = \tfrac{1}{\lambda}).$$

## §13. FRACTIONAL POWERS OF h, OF $(s-\alpha)^{-1}$, AND OF $(s-\alpha)$

FRACTIONAL POWER $h^\lambda$. We already know (2.3), i.e.,

$$h^n = \left\{\frac{t^{n-1}}{(n-1)!}\right\} = \left\{\frac{t^{n-1}}{\Gamma(n)}\right\} \qquad (n = 1,2,\ldots).$$

We can extend $h^n$ to $h^\lambda$ with $\text{Re } \lambda > 0$ through

$$h^\lambda = \frac{h^{\lambda+1}}{h} = \frac{\{\Gamma(\lambda+1)^{-1}t^\lambda\}}{h} = s\{\Gamma(\lambda+1)^{-1}t^\lambda\} \in C_H, \tag{13.1}$$

because if $\lambda$ = an integer $n \geqq 1$, we have

$$h^n = sh^{n+1}.$$

We shall call $h^\lambda$ ($\text{Re } \lambda > 0$; $h^0 = I$) the *fractional power of* h.

PROPOSITION 9. For $f \in C$, we have

$$\{h^{\lambda}f(t)\} = \left\{\frac{1}{\Gamma(\lambda)}\int_0^t (t-u)^{\lambda-1}f(u)\,du\right\} \in C \qquad (\mathrm{Re}\ \lambda > 0). \tag{13.2}$$

PROOF: By (13.1), we have

$$\begin{aligned} h^{\lambda}f &= s\left\{\Gamma(\lambda+1)^{-1}\int_0^t (t-u)^{\lambda}f(u)\,du\right\} \\ &= \frac{\partial}{\partial t}\left\{\Gamma(\lambda+1)^{-1}\int_0^t (t-u)^{\lambda}f(u)\,du\right\} + \left\{\Gamma(\lambda+1)^{-1}\int_0^0 ((-u)^{\lambda}f(u)\,du\right\} \\ &= \Gamma(\lambda+1)^{-1}((t-u)^{\lambda}f(u))_{u=t} + \int_0^t \Gamma(\lambda+1)^{-1}\lambda(t-u)^{\lambda-1}f(u)\,du \\ &= \left\{\Gamma(\lambda)^{-1}\int_0^t (t-u)^{\lambda-1}f(u)\,du\right\}. \end{aligned}$$

because $\lambda(t-u)^{\lambda-1}f(u)$ is integrable with respect to $u$ *uniformly in the parameter* $t$ when $t$ is restricted to within a finite closed interval $[0,t_0]$.

REMARK 13.1. $h^{\lambda}f$ is called the *fractional integration* of $f \in C$.

EXAMPLE 13.1. By $\Gamma(1/2) = \sqrt{\pi}$ and $(t-u)^{1/2-1} = \frac{1}{\sqrt{t-u}}$,

$$(h^{1/2}f)(t) = \frac{1}{\sqrt{\pi}}\int_0^t \frac{1}{\sqrt{t-u}}f(u)\,du. \tag{13.3}$$

FRACTIONAL POWER OF $(s-\alpha)^{-1}$. We shall extend the natural number $n$ in

$$(s-\alpha)^{-n} = \left(\frac{I}{(s-\alpha)}\right)^n = \frac{1}{(s-\alpha)^n} = \left\{\frac{t^{n-1}}{\Gamma(n)}e^{\alpha t}\right\}$$

to $\lambda$ with $\mathrm{Re}\ \lambda > 0$ through

$$(s-\alpha)^{-\lambda} = s\left\{\int_0^t \Gamma(\lambda)^{-1}u^{\lambda-1}e^{\alpha u}du\right\} \in C_H, \tag{13.4}$$

because if $\lambda$ = an integer $n \geqq 1$, we have

$$(s-\alpha)^{-n} = sh\{\Gamma(n)^{-1}t^{n-1}e^{\alpha t}\}.$$

We shall call (13.4) the *fractional power* of $(s-\alpha)^{-1}$.

Similarly as in Proposition 9, we have

PROPOSITION 9'. For $f \in C$, we have

$$\{(s-\alpha)^{-\lambda}f(t)\} = \left\{\int_0^t \frac{(t-u)^{\lambda-1}}{\Gamma(\lambda)}e^{\alpha(t-u)}f(u)\,du\right\} \in C. \tag{13.5}$$

THEOREM 11.

$$(s-\alpha)^{-\lambda}(s-\alpha)^{-\mu} = (s-\alpha)^{-\lambda-\mu}, \qquad (\mathrm{Re}\ \lambda > 0,\ \mathrm{Re}\ \mu > 0). \tag{13.6}$$

PROOF: We have, by (13.4),

$$\begin{aligned}(s-\alpha)^{-\lambda}(s-\alpha)^{-\mu} &= s\left\{h\ \frac{t^{\lambda-1}}{\Gamma(\lambda)}\ e^{\alpha t}\right\}s\left\{h\ \frac{t^{\mu-1}}{\Gamma(\mu)}\ e^{\alpha t}\right\}\\ &= s^2\left\{h\ \frac{t^{\lambda-1}}{\Gamma(\lambda)}\ e^{\alpha t}\right\}\left\{h\ \frac{t^{\mu-1}}{\Gamma(\mu)}\ e^{\alpha t}\right\}\\ &= s^2h^2\left\{\int_0^t \frac{(t-u)^{\lambda-1}e^{\alpha(t-u)}}{\Gamma(\lambda)\Gamma(\mu)}\ u^{\mu-1}e^{\alpha u}du\right\}\\ &= \frac{1}{\Gamma(\lambda)\Gamma(\mu)}\left\{e^{\alpha t}\int_0^t (t-u)^{\lambda-1}u^{\mu-1}du\right\}.\end{aligned}$$

By the substitution $t-u = tw$, we obtain

$$= \frac{-1}{\Gamma(\lambda)\Gamma(\mu)}\left\{\int_1^0 (tw)^{\lambda-1}(t-tw)^{\mu-1}te^{\alpha t}dw\right\}$$

$$= \frac{1}{\Gamma(\lambda)\Gamma(\mu)}\ \{t^{\lambda+\mu-1}e^{\alpha t}\}\int_0^1 w^{\lambda-1}(1-w)^{\mu-1}dw$$

$$= \left\{\frac{t^{\lambda+\mu-1}}{\Gamma(\lambda+\mu)}\ e^{\alpha t}\right\} \qquad \text{(by (12.5))}$$

$$= (s-\alpha)^{-\lambda-\mu} \qquad \text{(by (13.4))}.$$

COROLLARY. When $\alpha = 0$, we have

$$s^{-\lambda} = \left(\frac{I}{s}\right)^{\lambda} = \left(\frac{1}{s}\right)^{\lambda} = h^{\lambda} \qquad (\mathrm{Re}\ \lambda > 0) \tag{13.7}$$

so that

$$h^{\lambda}h^{\mu} = h^{\lambda+\mu} \qquad (\mathrm{Re}\ \lambda > 0,\ \ \mathrm{Re}\ \mu > 0;\ \ h^0 = I). \tag{13.8}$$

PROOF: By (13.4) and (13.1),

$$s^{-\lambda} = s\left\{\int_0^t \Gamma(\lambda)^{-1}u^{\lambda-1}du\right\} = s\{\Gamma(\lambda+1)^{-1}t^{\lambda}\} = h^{\lambda}.$$

FRACTIONAL POWER $(s-\alpha)^{\lambda}$ $(\mathrm{Re}\ \lambda > 0)$. We shall prove

PROPOSITION 10. For $0 < \mathrm{Re}\ \lambda < 1$, we define

$$(s-\alpha)^{\lambda} = (s-\alpha)s\left\{\int_0^t \Gamma(1-\lambda)^{-1}e^{\alpha u}u^{-\lambda}du\right\} \in C_H. \tag{13.9}$$

Then we have

$$(s-\alpha)^{\lambda}(s-\alpha)^{-\lambda} = I \qquad (0 < \mathrm{Re}\ \lambda < 1). \tag{13.10}$$

By (13.9) and (13.4),

$$(s-\alpha)^{\lambda}(s-\alpha)^{-\lambda} = \frac{(s-\alpha)}{\Gamma(1-\lambda)\Gamma(\lambda)}\left\{\int_0^t e^{\alpha(t-u)}(t-u)^{-\lambda}u^{\lambda-1}e^{\alpha u}du\right\}.$$

By the substitution $t-u = tw$,

$$(s-\alpha)^{\lambda}(s-\alpha)^{-\lambda} = \frac{(s-\alpha)\{e^{\alpha t}\}}{\Gamma(1-\lambda)\Gamma(\lambda)}\int_0^1 t^{-\lambda}w^{-\lambda}t^{\lambda-1}(1-w)^{\lambda-1}t\ dw$$

$$= \frac{I}{\Gamma(1-\lambda)\Gamma(\lambda)}B(1-\lambda,\lambda) = \frac{\Gamma(1-\lambda)\Gamma(\lambda)}{\Gamma(1-\lambda)\Gamma(\lambda)}\frac{I}{\Gamma(1)} = I.$$

REMARK 13.1. We sometimes call $s^{\lambda}f$ $(0 < \mathrm{Re}\ \lambda < 1)$ the *fractional differentiation* of $f \in C$.

EXAMPLE 13.2. We have

$$\frac{I}{s\sqrt{s+\alpha}} = \left\{\frac{2}{\sqrt{\alpha\pi}}\int_0^{\sqrt{\alpha t}} e^{-u^2}du\right\} \qquad (\alpha > 0). \tag{13.11}$$

PROOF: Since

$$\frac{I}{\sqrt{s+\alpha}} = (s+\alpha)^{-1/2} = s\left\{h\ \frac{t^{1/2-1}}{\Gamma(1/2)}\ e^{-\alpha t}\right\},$$

we have

$$\frac{I}{s\sqrt{s+\alpha}} = h\ \frac{I}{\sqrt{s+\alpha}} = \left\{\int_0^t \frac{u^{-1/2}}{\sqrt{\pi}}\ e^{-\alpha u}du\right\}$$

$$= \left\{\int_0^{\sqrt{\alpha t}} \frac{1}{\sqrt{\pi}}\left(\frac{v}{\sqrt{\alpha}}\right)^{-1} e^{-v^2}\cdot 2\ \frac{v}{\alpha}\ dv\right\}$$

$$= \left\{\frac{1}{\sqrt{\alpha}}\ \mathrm{Erf}\ \sqrt{\alpha t}\right\},$$

where

$$\mathrm{Erf}\ t = \mathrm{Erf}(t) = \frac{2}{\sqrt{\pi}}\int_0^t e^{-v^2}dv. \tag{13.12}$$

This function Erf(t) often occurs in the theory of probability and is termed the *error function*.

EXERCISES FOR §13. Prove the equality

$$\frac{I}{\sqrt{s}(s-\alpha)} = \left\{\frac{e^{\alpha t}}{\sqrt{\alpha}} \operatorname{Erf} \sqrt{\alpha t}\right\}. \tag{13.13}$$

GENERAL POWER $h^\gamma$ ($\gamma$ IS A COMPLEX NUMBER). We can define the *general power* $h^\gamma$ as follows:

$$\begin{cases} h^\gamma = \dfrac{h^{\gamma+n}}{h^n} = \dfrac{\{\Gamma(\gamma+n)^{-1}t^{\gamma+n-1}\}^*}{\{\Gamma(n)^{-1}t^{n-1}\}}, \\ n \text{ being any integer } \geqq 1 \text{ such that } \operatorname{Re}(\gamma+n) > 1. \end{cases} \tag{13.14}$$

REMARK 13.2. The above definition (13.14) is reasonable, i.e., it is without contradiction. This means that if we take another integer $m \geqq 1$ such that $\operatorname{Re}(\gamma+m) > 1$, then

$$h^m h^{\gamma+n} = \{\Gamma(\gamma+n+m)^{-1}t^{\gamma+n+m-1}\} = h^n h^{\gamma+m}. \tag{13.15}$$

PROOF: By the substitution $t-u = tw$, we obtain

$$\begin{aligned} h^m h^{\gamma+n} &= \left\{\Gamma(m)^{-1}\Gamma(\gamma+n)^{-1} \int_0^t (t-u)^{m-1}u^{\gamma+n-1}du\right\} \\ &= \left\{\Gamma(m)^{-1}\Gamma(\gamma+n)^{-1} \int_0^1 t^{m-1}w^{m-1}t^{\gamma+n-1}(1-w)^{\gamma+n-1}t\,dw\right\} \\ &= \{\Gamma(m)^{-1}\Gamma(\gamma+n)^{-1}B(m,\gamma+n)t^{\gamma+n+m-1}\} \\ &= \{\Gamma(\gamma+n+m)^{-1}t^{\gamma+n+m-1}\} \quad \text{(by (12.5))}. \end{aligned}$$

We shall further discuss the general power $s^\gamma$ and the general power $(s-\alpha)^\gamma$ in §19.

---

*Here $t^{\gamma+n-1} = \exp((\gamma+n-1)\log t)$ by taking the principal value of $\log t$ for $t > 0$.

# Chapter V
# Hyperfunctions Represented by Infinite Power Series in h

## §14. THE BINOMIAL THEOREM

THE DEFINITION OF $(1+z)^\alpha$. Let $\alpha$ be any complex number. For any complex number $z$ with $|z| < 1$, we define

$$\left\{\begin{array}{l}(1+z)^\alpha = e^{\alpha\log(1+z)}, \text{ where we take the branch of the} \\ \text{function } \log(1+z) \text{ determined by } \log(1+0) = 0.\end{array}\right. \tag{14.1}$$

Then we have

$$\frac{d}{dz}(1+z)^\alpha = e^{\alpha\log(1+z)}\frac{\alpha}{1+z} = e^{(\alpha-1)\log(1+z)}\alpha = \alpha(1+z)^{\alpha-1}. \tag{14.2}$$

We can prove

THEOREM 12.

$$\left\{\begin{array}{l}(1+z)^\alpha = \sum_{k=0}^{\infty}\binom{\alpha}{k}z^k, \text{ where } \binom{\alpha}{k} = \dfrac{\Gamma(\alpha+1)}{\Gamma(\alpha-k+1)k!} \\ \qquad = \dfrac{\alpha(\alpha-1)\cdots(\alpha-k+1)}{k!} \qquad (k = 1,2,3,\ldots;\ \binom{\alpha}{0} = 1),\end{array}\right. \tag{14.3}$$

and the series $\Sigma_{k=0}^{\infty}\binom{\alpha}{k}z^k$ is convergent for $|z| < 1$.

PROOF: $\log(1+z)$, determined by $\log(1+0) = 0$, is analytic for $|z| < 1$. Therefore, the Taylor series $\Sigma_{k=0}^{\infty}\alpha_k z^k$ of the function $(1+z)^\alpha$ is convergent for $|z| < 1$ and represents the function $(1+z)^\alpha$. The coefficients $\alpha_k$ are given by differentiation:

$$\alpha_k = \frac{1}{k!}\left(\frac{d^k}{dz^k}(1+z)^\alpha\right)_{z=0} \qquad (k = 1,2,\ldots;\ \alpha_0 = 1),$$

so that, by (14.2),

$$\alpha_1 = \frac{1}{1!}(\alpha(1+z)^{\alpha-1})_{z=0} = \frac{\alpha}{1!},$$

$$\alpha_2 = \frac{1}{2!}(\alpha(\alpha-1)(1+z)^{\alpha-2})_{z=0} = \frac{\alpha(\alpha-1)}{2!},$$

and so on.

EXAMPLES OF BINOMIAL EXPANSIONS (SERIES)

$$\frac{1}{1+t} = 1 - t + t^2 - t^2 + \cdots$$

$$\frac{1}{(1+t)^2} = 1 - 2t + 3t^2 - 4t^3 + \cdots$$

$$\sqrt{1+t} = 1 + \frac{1}{2}t - \frac{1}{2\cdot 4}t^2 + \frac{1\cdot 3}{2\cdot 4\cdot 6}t^3 - \frac{1\cdot 3\cdot 5}{2\cdot 4\cdot 6\cdot 8}t^4 + \cdots$$

$$\frac{1}{\sqrt{1+t}} = 1 - \frac{1}{2}t + \frac{1\cdot 3}{2\cdot 4}t^2 - \frac{1\cdot 3\cdot 5}{2\cdot 4\cdot 6}t^3 + \frac{1\cdot 3\cdot 5\cdot 7}{2\cdot 4\cdot 6\cdot 8}t^4 - \cdots .$$

## §15. BESSEL'S FUNCTION $J_n(t)$

THEOREM 13. The *Bessel function* $J_n(t)$ *of order* $n$ is defined as a solution $y(t)$ of the *Bessel** *differential equation*

$$y''(t) + \frac{1}{t}y'(t) + \left(1 - \frac{n^2}{t^2}\right)y(t) = 0, \tag{15.1}$$

and it can be obtained as a solution of (15.1) having the power series expansion

$$y(t) = C_0t^n + C_2t^{n+2} + C_4t^{n+4} + \cdots + C_{2\mu}t^{n+2\mu} + \cdots,$$
$$C_0 = \frac{1}{2^n\Gamma(n+1)}, \tag{15.2}$$

so that $y(t) = J_n(t)$:

$$J_n(t) = \sum_{\mu=0}^{\infty} \frac{(-1)^\mu}{\Gamma(n+\mu+1)\Gamma(\mu+1)}\left(\frac{t}{2}\right)^{n+2\mu} \quad (n = 0,1,2,\ldots). \tag{15.3}$$

PROOF: By the formal term-wise differentiation of (15.2), we have

$$y'' = n(n-1)C_0t^{n-2} + (n+2)(n+1)C_2y^n + \cdots +$$
$$(n+2\mu)(n+2\mu-1)C_{2\mu}t^{n+2\mu-2} + \cdots$$

$$t^{-1}y' = nC_0t^{n-2} + (n+2)C_2t^n + \cdots + (n+2\mu)C_{2\mu}t^{n+2\mu-2} + \cdots$$

* Friedrich Wilhelm Bessel (1784-1846).

$$y = C_0t^n + C_2t^{n+2} + \cdots + C_{2\mu}t^{n+2\mu} + \cdots$$

$$-\frac{n^2}{t^2}y = -n^2C_0t^{n-2} - n^2C_2t^n - \cdots - n^2C_{2\mu}t^{n+2\mu-2} - \cdots$$

Hence, by putting this formal power series equal to zero, we obtain

$$((n+2\mu)^2 - n^2)C_{2\mu} + C_{2(\mu-1)} = 0 \qquad (\mu = 0,1,2,\ldots), \tag{15.4}$$

since this is the coefficient of $t^{n+2\mu-2}$ equated with $0$. Thus

$$C_{2\mu} = \frac{-1}{(2n+2\mu)2\mu}C_{2(\mu-1)}$$

and so, combined with $C_0 = \dfrac{1}{2n\cdot\Gamma(n+1)}$, we obtain (15.3).

We next show that the formal power series given by (15.3) is absolutely convergent for every $t$. To prove it, let $k_0 > 1$ be a positive integer such that

$$\frac{t}{2k_0} = \delta < \frac{1}{2}.$$

Then, for any integer $k > k_0$, we have

$$\left|\frac{(-1)^k}{\Gamma(n+k+1)\Gamma(k+1)}\left(\frac{t}{2}\right)^{n+2k}\right|$$

$$\leqq \sum_{\mu=1}^{k_0-1}\left|\frac{1}{(n+\mu)}\frac{t}{2}\right| \cdot \left|\frac{1}{\mu}\frac{t}{2}\right| \times \left|\frac{t}{2}\right|^n \times \delta^{2(k-k_0+1)}.$$

Hence

$$\sum_{\mu=k_0}^{\infty}\left|\frac{(-1)^\mu}{\Gamma(n+\mu+1)\Gamma(\mu+1)}\left(\frac{t}{2}\right)^{n+2\mu}\right|$$

$$\leqq \left|\frac{t}{2}\right|^n \prod_{\mu=1}^{k_0-1} \left|\frac{1}{n+\mu}\frac{t}{2}\right| \cdot \left|\frac{1}{\mu}\frac{t}{2}\right| \times \left(\sum_{k=k_0}^{\infty} \delta^{2(k-k_0+1)}\right).$$

The right hand side is convergent since $\Sigma_{k=k_0}^{\infty}\ \delta^{2(k-k_0+1)}$ is a geometric series with $0 < \delta < 1/2$.

Finally we recall for the reader the well known

PROPOSITION 11. If a power series $p(t) = \Sigma_{n=0}^{\infty}\ \alpha_n t^n$ in $t$ is absolutely convergent at a point $t = t_0 \neq 0$, then $p(t)$ is continuously differentiable at any $t$ with $|t| < |t_0|$ and the derivative $p'(t)$ is obtained by termwise differentiation:

$$p'(t) = \sum_{k=0}^{\infty} k\alpha_k t^{k-1} \quad \text{at every } t \text{ with } |t| < |t_0|. \tag{15.5}$$

Therefore, the above method of formal term-wise differentiation is legitimate to obtain the Bessel function $J_n(t)$. We will see in §20 that $J_n(t)$ can be obtained by virtue of the Operational Calculus.

## §16. HYPERFUNCTIONS REPRESENTED BY POWER SERIES IN h

THEOREM 14. Let

$$\alpha_0 + \alpha_1 z + \alpha_2 z^2 + \cdots + \alpha_n z^n + \cdots \tag{16.1}$$

be a power series in complex variable $z$ with complex coefficients $\alpha_0, \alpha_1, \ldots$ . Assume that (16.1) is convergent at a certain value $z_0 \neq 0$ of $z$. Then the power series in $h$:

$$\alpha_0 h^0 + \alpha_1 h + \alpha_2 h^2 + \cdots + \alpha_n h^n + \cdots \tag{16.2}$$

defines, by $h^n = \Gamma(n)^{-1} t^{n-1}$, a hyperfunction

$$\alpha_0 I + \left(\alpha_1\{1\} + \alpha_2 t + \cdots + \alpha_n \frac{t^{n-1}}{(n-1)!} + \cdots\right) \in C_H. \tag{16.3}$$

PROOF: Since (16.1) is convergent at $z = z_0 \neq 0$, there exists a positive number $M < \infty$ such that

$$\sup_{n \geqq 0} |\alpha_n z_0^{\,n}| = M < \infty. \tag{16.4}$$

Thus

$$|\alpha_n| \leqq \frac{M}{|z_0|^n} \qquad (n = 0,1,2,\ldots).$$

Hence, for any $t_0 > 0$, there exists a natural number $k_0$ such that

$$\frac{t}{|z_0|} / k_0 = \delta < \frac{1}{2} \quad \text{for} \quad 0 \leqq t \leqq t_0.$$

Then we have, for $(n-1) \geqq k_0$,

$$\left|\alpha_n \frac{t^{n-1}}{(n-1)!}\right| \leqq \frac{|\alpha_n z_0^{\,n}|}{(n-1)!\,|z_0|} \left|\frac{t}{z_0}\right|^{n-1}$$

$$\leqq M \prod_{k=1}^{k_0-1} \left(\frac{1}{k} \frac{t}{|z_0|}\right) \frac{1}{|z_0|} \times \delta^{n-k_0} \leqq M \prod_{k=1}^{k_0-1} \left(\frac{1}{k} \frac{t}{|z_0|}\right) \frac{1}{|z_0|} \left(\frac{1}{2}\right)^{n-k_0}.$$

Hence

$$\sum_{n=k_0}^{\infty} \left|\alpha_n \frac{t^{n-1}}{(n-1)!}\right|$$

is uniformly convergent for $0 \leqq t \leqq t_0$.

Therefore,

$$\left\{\alpha_1\{1\} + \alpha_2 t + \ldots + \alpha_n \frac{t^{n-1}}{(n-1)!} + \ldots\right\}$$

belongs to $C[0,\infty)$, and so (16.2) belongs to $C_H$.

THEOREM 15. If $\alpha$ is a positive number, we have

$$\frac{I}{\sqrt{s^2+\alpha^2}} = \{J_0(\alpha t)\} \tag{16.5}$$

PROOF: We have

$$\frac{I}{s^2+\alpha^2} = \frac{h^2}{h^2(s^2+\alpha^2)} = \frac{h^2}{I+(\alpha h)^2}$$

and hence

$$\frac{I}{\sqrt{s^2+\alpha^2}} = \sqrt{\frac{I}{s^2+\alpha^2}} = \sqrt{\frac{h^2}{I+(\alpha h)^2}} = \frac{h}{\sqrt{I+(\alpha h)^2}} = h(I + (\alpha h)^2)^{-1/2}.$$

Here the equality

$$\frac{I}{\sqrt{I+(\alpha h)^2}} = (I + (\alpha h)^2)^{-1/2}$$

is seen from

$$\frac{1}{(1+(\alpha z)^2)^{1/2}} = (1 + (\alpha z)^2)^{-1/2}.$$

On the other hand, we have

$$\begin{aligned}
h(I + (\alpha h)^2)^{-1/2} &= h\sum_{k=0}^{\infty} \binom{-1/2}{k}(\alpha h)^{2k} \\
&= h\sum_{k=0}^{\infty} \frac{-\frac{1}{2}(-\frac{1}{2}-1)(-\frac{1}{2}-2)\cdots(-\frac{1}{2}-k+1)}{k!}\alpha^{2k}h^{2k} \\
&= \sum_{k=0}^{\infty} \frac{(-1)^k(2k)!}{2^{2k}(k!)^2}\alpha^{2k}h^{2k+1} \\
&= \left\{\sum_{k=0}^{\infty} \frac{(-1)^k(2k)!}{2^{2k}(k!)^2}\alpha^{2k}\frac{t^{2k}}{(2k)!}\right\} \\
&= \left\{\sum_{k=0}^{\infty} (-1)^k \frac{(\alpha t)^{2k}}{2^{2k}(k!)^2}\right\} = \{J_0(\alpha t)\}.
\end{aligned}$$

EXAMPLE 16.1.

$$\frac{1}{1-z} = 1 + z + z^2 + \cdots$$

is convergent for $|z| < 1$. Hence we have

$$\frac{I}{s-\alpha} = \frac{h}{I-\alpha h} = h + \alpha h^2 + \alpha^2 h^3 + \cdots$$

$$= \left\{1 + \frac{\alpha t}{1!} + \frac{\alpha^2 t^2}{2!} + \cdots\right\} = \{e^{\alpha t}\}.$$

EXAMPLE 16.2.

$$\sin z = z - \frac{z^3}{3!} + \frac{z^5}{5!} - \cdots$$

converges at every $z$. Hence

$$\sin\frac{I}{s} = \sin h = h - \frac{h^3}{3!} + \frac{h^5}{5!} - \cdots = \left\{1 - \frac{t^2}{3!2!} + \frac{t^4}{5!4!} - \cdots\right\}. \tag{16.6}$$

EXAMPLE 16.3. Let $\alpha$ be a positive number. then

$$(\sqrt{s^2+\alpha^2} - s)^n = \left\{\frac{n\alpha}{t} J_n(\alpha t)\right\} \quad (n = 1,2,\ldots). \tag{16.7}$$

PROOF: We shall make use of

$$(\sqrt{1+z} - 1)^n = n \sum_{k=0}^{\infty} (-1)^k \frac{(n+2k-1)!}{2^{n+2k} k!(n+k)!} z^{n+k}, \quad |z| < 1. \tag{16.8}$$

Prior to the proof of (16.8), we shall derive (16.7) from (16.8). Thus, by Theorem 13,

$$(\sqrt{s^2+\alpha^2} - s)^n = s^n(\sqrt{I + \alpha^2 h^2} - I)^n$$

$$= n \sum_{k=0}^{\infty} (-1)^k \frac{(n+2k-1)!}{2^{n+2k} k!(n+k)!} \alpha^{2(n+k)} h^{n+2k}$$

$$= \left\{n \sum_{k=0}^{\infty} (-1)^k \frac{\alpha^{2(n+k)} t^{n+2k-1}}{2^{n+2k} k!(n+k)!}\right\}$$

$$= \left\{\frac{n\alpha^n}{t} J_n(\alpha t)\right\}.$$

Hence, in particular,

$$(\sqrt{s^2+\alpha^2} - s)^n = \left\{n \sum_{k=0}^{\infty} (-1)^k \frac{\alpha^{2(n+k)} t^{n+2k-1}}{2^{n+2k} k!(n+k)!}\right\}. \tag{16.8$'$}$$

Now we shall prove (16.8). Put

$$f_n(z) = (\sqrt{1+z} - 1)^n \qquad (n = 1,2,\ldots). \tag{16.9}$$

Then it is easy to see

$$\begin{cases} f'_{n+1}(z) = \dfrac{(n+1)(\sqrt{1+z} - 1)^n}{2\sqrt{1+z}} \\[2ex] (z^{-n/2} f_n(z))' = \dfrac{n}{2} z^{-n/2-1} \dfrac{(\sqrt{1+z} - 1)^n}{\sqrt{1+z}}. \end{cases}$$

Hence

$$f'_{n+1}(z) = \frac{n+1}{n} z^{n/2+1} (z^{-n/2} f_n(z))'.$$

This, combined with $f_{n+1}(0) = 0$, gives

$$f_{n+1}(z) = \frac{n+1}{n} \int_0^z z^{n/2+1} (z^{-n/2} f_n(z))' dz. \tag{16.10}$$

On the other hand, we have

$$f_1(z) = \sqrt{1+z} - 1 = \sum_{\nu=0}^{\infty} (-1)^{\nu} \frac{(2\nu)!}{2^{1+2\nu} \nu! (1+\nu)!} z^{1+\nu}. \tag{16.11}$$

Therefore, by (16.10) and (16.11) we obtain (16.8).

COROLLARY OF (16.7). We have

$$\frac{(\sqrt{s^2+\alpha^2} - s)^n}{\sqrt{s^2+\alpha^2}} = \{\alpha^n J_n(\alpha t)\} \qquad (n = 0,1,2,\ldots). \tag{16.12}$$

PROOF: In (16.7) we substitute $(n+1)$ and then differentiate with respect to $\alpha$ and multiply by $((n+1)\alpha)^{-1}$.

EXAMPLE 16.4. Let $\alpha > 0$. Then

$$e^{-\alpha/s} = s\{J_0(2\sqrt{\alpha t})\}. \tag{16.13}$$

PROOF:

$$\begin{aligned} e^{-\alpha h} &= \sum_{k=0}^{\infty} (-1)^k \frac{(\alpha h)^k}{k!} = s \sum_{k=0}^{\infty} (-1)^k \frac{\alpha^k h^{k+1}}{k!} \\ &= s\left\{\sum_{k=0}^{\infty} (-1)^k \frac{\alpha^k t^k}{(k!)^2}\right\} = s\{J_0(2\sqrt{\alpha t})\}. \end{aligned}$$

EXAMPLE 16.5. Let $\alpha > 0$. Then

$$\frac{e^{-\alpha/s}}{\sqrt{s}} = \left\{\frac{1}{\sqrt{\pi t}} \cos 2\sqrt{\alpha t}\right\}. \tag{16.14}$$

PROOF:

$$h^{1/2}e^{-\alpha h} = \sum_{k=0}^{\infty} (-1)^k \frac{\alpha^k h^{k+1/2}}{k!} = \sum_{k=0}^{\infty} (-1)^k \frac{\alpha^k t^{k-1/2}}{k!\Gamma(k+1/2)}$$

$$= \sum_{k=0}^{\infty} (-1)^k \frac{\alpha^k t^{k-1/2}}{k(k-1)\cdots 1(k-1/2)(k-3/2)\cdots(1/2)\Gamma(1/2)}$$

$$= \sum_{k=0}^{\infty} (-1)^k \frac{2^{2k}\alpha^k t^{k-1/2}}{(2k)(2k-1)\cdots 1\cdot\sqrt{\pi}}$$

$$= \frac{1}{\sqrt{\pi t}} \sum_{k=0}^{\infty} (-1)^k \frac{(2\sqrt{\alpha t})^{2k}}{(2k)!}$$

$$= s\left\{\int_0^t \frac{1}{\sqrt{\pi u}} \cos 2\sqrt{\alpha u}\, du\right\}.$$

EXERCISES FOR §16. Prove 1-3 below.

1. $\frac{I}{s^2} e^{-\alpha/s} = \{\sqrt{t/\alpha}\, J_1(2\sqrt{\alpha t})\}$

2. $\frac{I}{\sqrt{s}} e^{\alpha/s} = \left\{\frac{1}{\sqrt{\pi t}} \cosh (2\sqrt{\alpha t})\right\}$

3. $\frac{I}{s} e^{-\alpha/s} = \{J_0(2\sqrt{\alpha t})\}$.

# Part II
# Linear Ordinary Differential Equations with Linear Coefficients (The Class C/C of Hyperfunctions)

## Chapter VI
## The Titchmarsh Convolution Theorem and the Class C/C

### §17. PROOF OF THE TITCHMARSH CONVOLUTION THEOREM*

THEOREM 16. (THE TITCHMARSH CONVOLUTION THEOREM). If the convolution $fg$ of two functions $f$ and $g \in C$ is 0, then either $f = 0$ or $g = 0$ must be true.

PROOF: THE FIRST STEP (THE CASE $f = g$). We shall start with

$$\int_0^t f(t-\tau)f(\tau)d\tau = 0 \quad \text{for all} \quad t \geqq 0. \tag{17.1}$$

Let $T > 0$ and $\beta$ be real numbers. Put

$$\hat{f}(t) = e^{-i\beta t}f(t), \text{ where } i = \sqrt{-1}. \tag{17.2}$$

Then, for any real number $\alpha$, we obtain

$$\begin{aligned}\left(\int_{-T}^{T} e^{\alpha u}\hat{f}(T-u)\,du\right)^2 &= \int_{-T}^{T}\int_{-T}^{T} e^{\alpha(u+v)}\hat{f}(T-u)\hat{f}(T-v)\,du\,dv \\ &= \int_{-T}^{T} dv\left(\int_{u=-v}^{T} e^{\alpha(u+v)}\hat{f}(T-u)\hat{f}(T-v)\,du\right) \\ &\qquad + \int_{-T}^{T} dv\left(\int_{-T}^{u=-v} e^{\alpha(u+v)}\hat{f}(T-u)\hat{f}(T-v)\,du\right) \\ &= I_{\alpha,\beta,1} + I_{\alpha,\beta,2}.\end{aligned} \tag{17.3}$$

By the substitution

---

*Kôsaku Yosida and Shigetake Matsuura: A note on Mikusiński's proof of the Titchmarsh convolution theorem, to be published in the Contemporary Mathematics Series of the Amer. Math. Society.

$u = T - (t-\tau)$, $v = T-\tau$ with

$2T \geqq u+v \geqq 0$, $T \geqq u \geqq -T$, $T \geqq v \geqq -T$ and

$\frac{\partial(u,v)}{\partial(t,\tau)} = 1$,

we have, by (17.1) and (17.2),

$$I_{\alpha,\beta,1} = \int_0^{2T} dt\left(\int_0^t e^{\alpha(2T-t)}\hat{f}(t-\tau)\hat{f}(\tau)d\tau\right)$$

$$= \int_0^{2T} e^{\alpha(2T-t)-i\beta t}\left(\int_0^t f(t-\tau)f(\tau)d\tau\right)dt \tag{17.4}$$

$$= 0.$$

If $\alpha > 0$, then $e^{\alpha(u+v)} \leqq 1$ in the integration domain of the integral $I_{\alpha,\beta,2}$. Hence

$$|I_{\alpha,\beta,2}| \leqq \int_{-T}^{T} dv\left(\int_{-T}^{u=-v} |\hat{f}(T-u)\hat{f}(T-v)|du\right)$$
$$\leqq \int_{-T}^{T}\int_{-T}^{T} |f(T-u)f(T-v)|dudv \leqq M^2 \quad \text{with} \quad 0 \leqq M < \infty. \tag{17.5}$$

Therefore, by (17.3), (17.4) and (17.5), we have

$$\left|\int_{-T}^{T} e^{(\alpha+i\beta)u}f(T-u)du\right| \leqq M \quad \text{whenever} \quad \alpha \geqq 0. \tag{17.6}$$

Thus the Laplace integral

$$F_T(z) = \int_0^T e^{zu}f(T-u)du \tag{17.7}$$

satisfies, for $\alpha = \text{Re } z \geqq 0$,

$$|F_T(z)| \leqq N \quad \text{with} \quad N = M + \int_{-T}^{0}|f(T-u)|du. \tag{17.8}$$

Hence, we easily prove that the entire analytic function (17.7) satisfies

$$|F_T(z)| = \left|\int_0^T e^{zu}f(T-u)du\right| \leqq N_1 \quad \text{for all} \quad z, \text{ where}$$
$$N_1 = \max(N,K) \quad \text{with} \quad K = \max_{0\leqq u\leqq T} T \times |f(T-u)|. \tag{17.9}$$

Therefore, by Liouville's theorem* in analytic function theory, we have

* Joseph Liouville (1809-1882).

$$F_T(z) = \text{a constant for all } z.$$

Thus, by differentiating $F_T(z)$ at $z = 0$, we obtain

$$\int_0^T u^n f(T-u)du = 0, \qquad (n = 1,2,\ldots),$$

and so, for every polynomial $p(u)$ in $u$, we have

$$\int_0^T p(u)uf(T-u)du = 0$$

Hence, by Weierstrass' polynomial approximation theorem*, we obtain

$$\int_0^T q(u)uf(T-u)du = 0$$

for every continuous function $q(u)$. This proves that $uf(T-u) = 0$ for $0 \leqq u \leqq T$, i.e., $f(t)$ must vanish for $0 \leqq t \leqq T$. As $T > 0$ was arbitrarily chosen, we see that $f(t) \equiv 0$.

THE SECOND STEP (THE GENERAL CASE $f \neq g$). We start with

$$(fg)(t) = \int_0^t f(t-\tau)g(\tau)d\tau = 0 \quad \text{for all } t \geqq 0. \tag{17.10}$$

We put, following J. Mikusiński [5], p. 22,

$$f_1(t) = tf(t) \quad \text{and} \quad g_1(t) = tg(t). \tag{17.11}$$

Then

$$f_1g + fg_1 = 0, \tag{17.12}$$

because

$$\int_0^t (t-\tau)f(t-\tau)g(\tau)d\tau + \int_0^t f(t-\tau)\tau g(\tau)d\tau$$
$$= t\int_0^t f(t-\tau)g(\tau)d\tau = 0 \quad \text{by (17.10)}.$$

Hence, by making use of Theorem 1 (§1) and (17.10), we obtain

$$0 = (fg_1)(f_1g + fg_1) = (fg)(f_1g_1) + (fg_1)(fg_1)$$
$$= (fg_1)(fg_1).$$

---

*Karl Theodor Wilhelm Weierstrass (1815-1897). The theorem reads: Let $q(t)$ be any continuous function on the closed interval $[0,T]$. Then, for any number $\varepsilon > 0$, there exists a polynomial $p(t)$ such that $\max_{0 \leqq t \leqq T} |q(t)-p(t)| < \varepsilon$. For a proof, see §21 of this book.

By virtue of the first step, we thus have $fg_1 = 0$, i.e.,

$$\int_0^t f(t-\tau)\tau g(\tau)d\tau \equiv 0. \tag{17.13}$$

Repeating the same reasoning, we have

$$\int_0^t f(t-\tau)\tau^n g(\tau)d\tau \equiv 0 \quad (n = 0,1,2,\ldots). \tag{17.14}$$

Hence, for every polynomial $p(t)$, we obtain

$$\int_0^t f(t-\tau)p(\tau)g(\tau)d\tau \equiv 0$$

and so, by Weierstrass' polynomial approximation theorem,

$$\int_0^t f(t-\tau)q(\tau)g(\tau)d\tau \equiv 0$$

for every continuous function $q(t)$.

This proves that

$$f(t-\tau)g(\tau) = 0 \quad \text{for} \quad 0 \leqq \tau \leqq t < \infty.$$

Therefore, either $f(t) \equiv 0$ or $g(t) \equiv 0$ must be true.

## §18. THE CLASS $C/C$ OF HYPERFUNCTIONS

The class $C_H$ of hyperfunctions was introduced on the basis of Proposition 3 (§3). Similarly, on the basis of Theorem 16 (§17), we can introduce the class $C/C$ of hyperfunctions. In fact, we can prove

PROPOSITION 12. If $f$ and $g$ are in $C = C[0,\infty)$ and $g \neq 0$, then we can define the "fraction" $\frac{f}{g} = f/g$ by the equivalence relation $\approx$:

$$\frac{f}{g} \approx \frac{f_1}{g_1} \quad \text{means} \quad fg_1 = f_1 g, \tag{18.1}$$

because we can prove

$$\frac{f}{g} \approx \frac{f}{g}, \tag*{(18.1)$_1$}$$

$$\frac{f}{g} \approx \frac{f_1}{g_1} \quad \text{implies} \quad \frac{f_1}{g_1} \approx \frac{f}{g}, \tag*{(18.1)$_2$}$$

$$\frac{f}{g} \approx \frac{f_1}{g_1}, \ \frac{f_1}{g_1} \approx \frac{f_2}{g_2} \quad \text{implies} \quad \frac{f}{g} \approx \frac{f_2}{g_2}. \tag*{(18.1)$_3$}$$

PROOF: We shall prove $(18.1)_3$ and omit the others. Thus we start with

$$fg_1 = f_1 g \quad \text{and} \quad f_1 g_2 = f_2 g_1.$$

Hence

$$fg_1 g_2 = f_1 g g_2 \quad \text{and} \quad f_1 g_2 g = f_2 g_1 g$$

and so

$$fg_1 g_2 = f_2 g_1 g, \quad \text{that is,} \quad g_1(fg_2 - f_2 g) = 0.$$

Since $g_1 \neq 0$ as the denominator of the fraction $\frac{f_1}{g_1}$, we obtain by Theorem 16,

$$fg_2 - f_2 g = 0.$$

This proves the equivalence $\frac{f}{g} \approx \frac{f_2}{g_2}$.

Therefore, we can write

$$\frac{f}{g} = \frac{f_1}{g_1} \quad \text{if and only if} \quad fg_1 = f_1 g \quad \text{and} \quad g \neq 0, \quad g_1 \neq 0. \qquad (18.1)'$$

The totality $C/C$ of fractions of the form $\frac{f}{g}$ constitutes a ring* with the *sum* and *product* given by

$$\frac{f}{g} + \frac{f'}{g'} = \frac{fg' + f'g}{gg'} \qquad (18.2)$$

$$\frac{f}{g}\frac{f'}{g'} = \frac{ff'}{gg'}. \qquad (18.3)$$

In the above, $gg' \neq 0$ by Theorem 16, since $g \neq 0$ and $g' \neq 0$. This ring $C/C$ is *commutative* with respect to the product.

REMARK 18.1. As consequences of (18.1)', (18.2), and (18.3), we have the following important results (i)-(iii):

(i) Fractions of the form $\frac{0}{g}$ $(g \neq 0)$ are mutually equal to one another, and they represent the *zero of the ring* $C/C$ so that

$$\frac{0}{g} + \frac{f'}{g'} = \frac{f'}{g'} \quad \text{and} \quad \frac{0}{g}\frac{f'}{g'} = \frac{0}{g}. \qquad (18.4)$$

(ii) Fractions of the form $\frac{g}{g}$ $(g \neq 0)$ are mutually equal to one another, and they represent the (*multiplicative*) *unit* of the ring $C/C$ so that

$$\frac{g}{g}\frac{f'}{g'} = \frac{f'}{g'} \qquad (18.5)$$

---

*Similarly as $C_H$.

(iii) A fraction of the form $\frac{f}{g}$ has a (*multiplicative*) *inverse* if and only if $f \neq 0$, and, in this case,

$$\frac{f}{g}\frac{g}{f} = I, \quad \text{i.e.,} \quad I/(\frac{f}{g}) = (\frac{f}{g})^{-1} = \frac{g}{f}. \tag{18.6}$$

Hence, in case $f \neq 0$, $\frac{f}{g}$ and $\frac{g}{f}$ are (*multiplicative*) *inverses* of each other.

REMARK 18.2. Since $\frac{f}{h^n}$ is considered as a fraction in $C/C$, the ring $C_H$ is a *subring* of the ring $C/C$. Thus we may call fractions in $C/C$ hyperfunctions. In this way, the notion of operator or hyperfunction given in Part I is generalized to a much bigger extent by $C/C$.

# Chapter VII
# The Algebraic Derivative Applied to Laplace's Differential Equation

## §19. THE ALGEBRAIC DERIVATIVE

Pierre Simon Laplace (1749 - 1827) in his treatise "Théorie analytique des probabilités" of 1817 considered a differential equation which now carries his name and which may be written as

$$(a_2t+b_2)y''(t) + (a_1t+b_1)y'(t) + (a_0t+b_0)y(t) = 0, \tag{19.1}$$

where the a's and b's are given complex numbers with $a_2 \neq 0$.

Laplace's method of integration of (19.1) was to use a transformation

$$y(t) \longrightarrow Y(\sigma) = \int_0^\infty e^{-\sigma t}\, y(t)dt, \tag{19.2}$$

which is now known as the *Laplace transform.*

As we have discussed in Chapter III, we could solve linear ordinary differential equations with constant coefficients *algebraically* by making use of hyperfunctions of the form

$$\frac{q(s)}{p(s)} \quad (p(s) \text{ and } q(s) \text{ are polynomials in } s)$$

belonging to the ring $C_H$.

In view of the importance of the equation (19.1) in analysis, the purpose of the next section, §20, is to show that we can solve (19.1) *algebraiclly* by making use of the hyperfunctions

$$(s-\alpha I)^\gamma \quad (\alpha \text{ and } \gamma \text{ are complex numbers})$$

belonging to the ring $C/C$.

To this aim, we first explain the notion of the *algebraic derivative* D due to J. Mikusiński [5], and then we shall develop the *calculus of* D* when applied to the fractional powers $h^\gamma$, $(I-\alpha h)^\gamma$ and $(s-\alpha I)^\gamma$.

DEFINITION 19.1. The algebraic derivative D is a mapping of $C/C$ into $C/C$ such that

$$\begin{cases} Df = \{-tf(t)\} & \text{for } f \in C, \\ D\dfrac{f}{g} = \dfrac{(Df)g - f(Dg)}{g^2} & \text{for } \dfrac{f}{g} \in C/C. \end{cases} \tag{19.3}$$

PROPOSITION 13. Definition (19.3) is consistent; that is,

$$Df = D\frac{fg}{g} = \frac{(D(fg))g - (fg)Dg}{g^2} \tag{19.4}$$

$$\frac{f}{g} = \frac{f_1}{g_1} \text{ implies } D\frac{f}{g} = D\frac{f_1}{g_1}. \tag{19.5}$$

PROOF: We have

$$D(ff_1) = (Df)f_1 + f(Df_1), \tag{19.6}$$

because

$$\begin{aligned} D(ff_1) &= \left\{-t\int_0^t f(t-u)f_1(u)du\right\} \\ &= \left\{-\int_0^t (t-u)f(t-u)f_1(u)du + \int_0^t f(t-u)(-u)f_1(u)du\right\} \\ &= (Df)f_1 + f(Df_1). \end{aligned}$$

Hence,

$$\begin{aligned} D\frac{fg}{g} &= \frac{(D(fg))g - fg(Dg)}{g^2} = \frac{((Df)g + f(Dg))g - fg(Dg)}{g^2} \\ &= \frac{(Df)g}{g} = Df. \end{aligned}$$

Next we have, by the commutativity of the ring $C$ and (19.3)-(19.4),

$$\begin{aligned} D\frac{fg_1}{gg_1} &= \frac{(D(fg_1))gg_1 - fg_1(D(gg_1))}{(gg_1)^2} \\ &= \frac{(Df)g_1gg_1 + f(Dg_1)gg_1 - fg_1(Dg)g_1 - fg_1g(Dg_1)}{(gg_1)^2} \end{aligned}$$

*K. Yosida [10]: The algebraic derivative and Laplace's differential equation, Proc. Japan Acad., *59*, Ser. A, No. 1 (1983), 1-4.

$$= \frac{((Df)g - f(Dg))g_1g_1 + fg_1((Dg_1)g - g(Dg_1))}{g^2g_1^2}$$

$$= \frac{(Df)g - f(Dg)}{g^2} = D\,\frac{f}{g}.$$

Similarly we have

$$D\,\frac{f_1}{g_1}\frac{g}{g} = D\,\frac{f_1}{g_1}\,.$$

On the other hand,

$$\begin{cases} \dfrac{f}{g} = \dfrac{f_1}{g_1} \quad \text{implies} \quad fg_1 = gf_1 \quad \text{and so} \\ \dfrac{f}{g} = \dfrac{fg_1}{gg_1} = \dfrac{gf_1}{gg_1}, \end{cases}$$

proving that

$$D\,\frac{f}{g} = D\,\frac{fg_1}{gg_1} = D\,\frac{gf_1}{gg_1} = D\,\frac{f_1}{g_1}\,.$$

PROPOSITION 14. We have

$$D([\alpha]\,\frac{f}{g}] = [\alpha](D\,\frac{f}{g}), \tag{19.7}$$

$$D\Big(\frac{f}{g} + \frac{f_1}{g_1}\Big) = D\,\frac{f}{g} + D\,\frac{f_1}{g_1}\,, \tag{19.8}$$

$$D\Big(\frac{f}{g}\,\frac{f_1}{g_1}\Big) = (D\,\frac{f}{g})\frac{f_1}{g_1} + \frac{f}{g}\Big(D\,\frac{f_1}{g_1}\Big). \tag{19.9}$$

PROOF: We have

$$D[\alpha] = 0; \quad \text{in particular, } DI = 0, \tag{19.10}$$

because

$$D[\alpha] = D\,\frac{\{\alpha\}}{h} = \frac{(D\{\alpha\})h - \{\alpha\}(Dh)}{h^2}$$

$$= \frac{\{-t\alpha\}h - \{\alpha\}\{-t\}}{h^2} = \frac{\{-\alpha t^2/2\} + \{\alpha t^2/2\}}{h^2} = 0.$$

We next prove (19.8). We have, for $f$ and $f_1 \in C$,

$$\begin{aligned} D(f + f_1) &= \{-t(f(t) + f_1(t))\} \\ &= \{-tf(t)\} + \{-tf_1(t)\} = Df + Df_1. \end{aligned} \tag{19.11}$$

Since

$$\frac{f}{g} + \frac{f_1}{g_1} = \frac{fg_1}{gg_1} + \frac{f_1 g_1}{gg_1},$$

we may restrict ourselves, in the proof of (19.8), to the case $g = g_1 = k$. Thus

$$\begin{aligned} D\,\frac{f+f'}{k} &= \frac{(Df+Df')k - (Dk)(f+f')}{k^2} \\ &= \frac{(Df)k - (Dk)f}{k^2} + \frac{(Df')k - (Dk)f'}{k^2} \\ &= D\,\frac{f}{k} + D\,\frac{f'}{k}\,. \end{aligned}$$

We then prove (19.9) in the case $g = g_1 = k$. Thus

$$\begin{aligned} D\,\frac{f}{k}\,\frac{f_1}{k} &= D\,\frac{ff_1}{k^2} = \frac{(D(ff_1))k^2 - ff_1(Dk^2)}{k^4} \\ &= \frac{(Df)f_1k^2 + f(Df_1)k^2}{k^4} - \frac{2k(Dk)ff_1}{k^4} \\ &= \frac{(Df)k - (Dk)f}{k^2}\,\frac{f_1}{k} + \frac{f}{k}\,\frac{(Df_1)k - (Dk)f_1}{k^2} \\ &= (D\,\frac{f}{k})\frac{f_1}{k} + \frac{f}{k}\left(D\,\frac{f_1}{k}\right), \end{aligned}$$

and so, by (19.5),

$$\begin{aligned} D\left(\frac{f}{g}\,\frac{f_1}{g_1}\right) &= D\left(\frac{g_1}{g_1}\,\frac{f}{g}\,\frac{gf_1}{gg_1}\right) \\ &= \left(D\,\frac{g_1}{g_1}\,\frac{f}{g}\right)\frac{gf_1}{gg_1} + \frac{g_1}{g_1}\,\frac{f}{g}\left(D\,\frac{gf_1}{g_1g}\right). \end{aligned}$$

Hence (19.9) is proved.

Finally, by (19.10) and (19.9), we obtain (19.7). In fact,

$$D([\alpha]\,\frac{f}{g}) = (D[\alpha])\frac{f}{g} + [\alpha](D\,\frac{f}{g}) = [\alpha](D\,\frac{f}{g}).$$

PROPOSITION 15.

$$\begin{cases} \text{If } \ a = \frac{m}{n} \in C/C \ \text{ and } \ b = \frac{p}{q} \in C/C, \ \text{ then} \\ D\,\frac{a}{b} = \frac{(Da)b - a(Db)}{b^2} = D\,\frac{mq}{np}\,. \end{cases} \tag{19.12}$$

PROOF: We have

$$\frac{a}{b} = a\frac{I}{b} = \frac{m}{n}\frac{q}{p} = \frac{mq}{np} \quad \text{(by (18.6))}.$$

And

$$D\frac{mq}{np} = \frac{(D(mq))np - mq(D(np))}{n^2p^2}$$

$$= \frac{((Dm)q + m(Dq))np - mq((Dn)p + n(Dp)}{n^2p^2}$$

$$= \frac{((Dm)n - m(Dn))pq - mn((Dp)q - p(Dq))}{n^2p^2},$$

$$D\frac{a}{b} = \frac{(Da)b - a(Db)}{b^2}$$

$$= \left(\frac{(Dm)n - m(Dn))}{n^2}\frac{p}{q} - \frac{m}{n}\frac{(Dp)q - p(Dq)}{q^2}\right)/(\frac{p}{q})^2$$

$$= \frac{((Dm)n - m(Dn))pq - mn((Dp)q - p(Dq))}{n^2p^2}.$$

The preceding formulas are given, roughly, in J. Mikusiński [1]. We shall now prove

THEOREM 17. (i) For any complex number $\gamma$, we have

$$Dh^\gamma = -\gamma h^{\gamma+1}, \tag{19.13}$$

where, as was defined in (13.14),

$$\begin{cases} h^\gamma = \dfrac{h^{\gamma+n}}{h^n} = \dfrac{\{\Gamma(\gamma+n)^{-1}t^{\gamma+n-1}\}}{\{\Gamma(n)^{-1}t^{n-1}\}}, \\ n \text{ being any integer } \geqq 1 \text{ such that } \operatorname{Re}(\gamma+n) > 1. \end{cases} \tag{19.14}$$

(ii) For complex numbers $\alpha$ and $\gamma$, we have

$$D(I - \alpha h)^\gamma = \gamma(I - \alpha h)^{\gamma-1}\alpha h^2, \tag{19.15}$$

where

$$(I - \alpha h)^\gamma = \sum_{k=0}^{\infty}\binom{\gamma}{k}(-\alpha)^k h^k. \tag{19.16}$$

(iii) We have

$$D(s - \alpha I)^\gamma = \gamma(s - \alpha I)^{\gamma-1}, \quad \text{where} \tag{19.17}$$

$$(s - \alpha I)^{\gamma} = (s - \alpha)^{\gamma} = \frac{(I-\alpha h)^{\gamma}}{h^{\gamma}} . \tag{19.18}$$

PROOF: (19.13). We have

$$Dh^{n} = -nh^{n+1} \qquad (n = 0,1,2,\ldots;\ \ h^{0} = I), \tag{19.13)'$$

because

$$Dh^{n} = D\left(\frac{t^{n-1}}{(n-1)!}\right) = -\frac{t^{n}}{(n-1)!} = -\frac{n!h^{n+1}}{(n-1)!} = -nh^{n+1}.$$

In particular,

$$Dh^{0} = DI = 0 \quad \text{in accordance with (19.10).}$$

Hence we have, by (19.14) (= (13.14)) and (19.12),

$$\begin{aligned} Dh^{\gamma} &= D\,\frac{h^{\gamma+n}}{h^{n}} = \frac{(Dh^{\gamma+n})h^{n} - h^{\gamma+n}(Dh^{n})}{h^{2n}} \\ &= \frac{\{-t\Gamma(\gamma+n)^{-1}t^{\gamma+n-1}\}h^{n} - h^{\gamma+n}(-nh^{n+1})}{h^{2n}} \\ &= \frac{-(\Gamma(\gamma+n)^{-1}\Gamma(\gamma+n+1)h^{\gamma+n+1}h^{n} + nh^{\gamma+2n+1}}{h^{2n}} \\ &= \frac{-(\gamma+n-n)h^{\gamma+2n+1}}{h^{2n}} = -\gamma h^{\gamma+1}. \end{aligned}$$

(19.15). Since the binomial expansion

$$\sum_{k=0}^{\infty} \binom{\gamma}{k}(-\alpha z_{0})^{k}$$

converges for $|\alpha z_{0}| < 1$, the series

$$\sum_{k=1}^{\infty} \binom{\gamma}{k}(-\alpha)^{k}h^{k} = \sum_{k=1}^{\infty} \binom{\gamma}{k}(-\alpha)^{k}\,\frac{t^{k-1}}{(k-1)!}$$

converges for every $t$, thanks to the factors $((k-1)!)^{-1}$ (see Theorem 14 in §16). Hence, by $DI = 0$,

$$\begin{aligned} D(I - \alpha h)^{\gamma} &= DI + \left\{-t \sum_{k=1}^{\infty} \binom{\gamma}{k}(-\alpha)^{k}\Gamma(k)^{-1}t^{k-1}\right\} \\ &= \sum_{k=1}^{\infty} \binom{\gamma}{k}(-\alpha)^{k-1}\alpha kh^{k+1} \\ &= \gamma \sum_{k=1}^{\infty} \binom{\gamma-1}{k-1}(-\alpha)^{k-1}h^{k-1}\alpha h^{2} = \gamma(I-\alpha h)^{\gamma-1}\alpha h^{2}. \end{aligned}$$

(19.17). By (19.12), we have

$$D(s - \alpha I)^{\gamma} = D\,\frac{(I-\alpha h)^{\gamma}}{h^{\gamma}} = \frac{(D(I-\alpha h)^{\gamma})h^{\gamma} - (I-\alpha h)^{\gamma}(Dh^{\gamma})}{h^{2\gamma}}$$

$$= \frac{\gamma(I-\alpha h)^{\gamma-1}(\alpha h^2 h^{\gamma} + (I-\alpha h)h^{\gamma+1})}{h^{2\gamma}}$$

$$= \frac{\gamma(I-\alpha h)^{\gamma-1}h^{\gamma+1}}{h^{2\gamma}} = \gamma(s-\alpha I)^{\gamma-1}.$$

REMARK 19.1. Thus the algebraic derivative $D$ may be denoted by $\frac{d}{ds}$ when applied to "functions of $s$":

$$D = \frac{d}{ds}. \tag{19.19}$$

For example, we have

$$Ds^{\gamma} = \gamma s^{\gamma-1} \tag{19.17)'$$

and (19.17).

REMARK 19.2. As a corollary of (19.18), we have

$$(s-\alpha I)^{\gamma}(s-\alpha I)^{\delta} = (s-\alpha I)^{\gamma+\delta} \tag{19.20}$$

for any triple

$$\{\alpha,\gamma,\delta\}$$

of complex numbers $\alpha$, $\gamma$ and $\delta$. Thus (13.6) is now completely generalized, and so we have

$$s^{\gamma}s^{\delta} = s^{\gamma+\delta} \tag{19.20)'$$

and

$$h^{\gamma}h^{\delta} = h^{\gamma+\delta}. \tag{19.20)''$$

Thus (13.8) is completely generalized.

PROOF OF (19.20): By (14.1), we obtain

$$(I-\alpha z)^{\gamma}(I-\alpha z)^{\delta} = \exp(\gamma \log(1-\alpha z))\exp(\delta \log(1-\alpha z))$$

$$= \exp((\gamma+\delta)\log(1-\alpha z))$$

$$= (1-\alpha z)^{\gamma+\delta}.$$

This proves

$$(s-\alpha I)^{\gamma}(s-\alpha I)^{\delta} = \frac{(I-\alpha h)^{\gamma}}{h^{\gamma}} \frac{(I-\alpha h)^{\delta}}{h^{\delta}}$$
$$= \frac{(I-\alpha h)^{\gamma+\delta}}{h^{\gamma+\delta}} = (s-\alpha I)^{\gamma+\delta}$$

PROOF OF (19.20)": We have

$$s^{\gamma} = h^{-\gamma}, \text{ in particular } s^{0} = h^{0} = I \text{ and } Ds = I, \qquad (19.20)'''$$

by (19.18) and (19.17)'.

## §20. LAPLACE'S DIFFERENTIAL EQUATION

In the equation given by (19.1), we may assume that $b_2 = 0$ by taking

$$\left(t - \frac{b_2}{a_2}\right)$$

as a new independent variable. Thus we shall discuss the differential equation

$$\begin{cases} a_2 t y''(t) + (a_1 t + b_1) y'(t) + (a_0 t + b_0) y(t) = 0 \\ \text{with } a_2 \neq 0 \end{cases} \qquad (20.1)$$

in the domain $t \geqq 0$.

For *the twice continuously differentiable solution* $y(t)$ of (20.1), given initial conditions $y(0)$ and $y'(0)$, we obtain, by (5.6) and (19.3), the equation

$$-a_2 D(s^2 y - s[y(0)] - [y'(0)])$$
$$- a_1 D(sy - [y(0)]) + b_1(sy - [y(0)]) - a_0 Dy + b_0 y = 0.$$

This is, by (19.9), (19.10) and (19.17)', the same as

$$-(a_2 s^2 + a_1 s + a_0) Dy - (2a_2 s + a_1 - b_1 s - b_0) y$$
$$+ (a_2 - b_1)[y(0)] = 0. \qquad (20.2)$$

Therefore, we have

PROPOSITION 16. Assuming that the above solution $y(t)$ is $\not\equiv 0$ and satisfies the *initial condition*

$$y(0) = 0 \quad \text{if} \quad a_2 \neq b_1, \qquad (20.3)$$

we can convert (20.1) into

$$\frac{Dy}{y} = \frac{(-2a_2 + b_1)s - a_1 + b_0}{a_2 s^2 + a_1 s + a_0 I}. \tag{20.4}$$

We thus have

THEOREM 18. If the algebraic equation

$$a_2 z^2 + a_1 z + a_0 = 0 \tag{20.5}$$

has two distinct roots $z_1$ and $z_2$, then (20.4) is solvable by a hyperfunction $y \in C/C$:

$$y = C(s-z_1 I)^{\gamma_1}(s-z_2 I)^{\gamma_2} \quad (C \text{ is a non-zero constant}) \tag{20.6}$$

where complex numbers $\gamma_1$ and $\gamma_2$ are given from the partial fraction expansion

$$\frac{(-2a_2+b_1)s - a_1 + b_0}{a_2 s^2 + a_1 s + a_0 I} = \frac{\gamma_1}{s - z_1 I} + \frac{\gamma_2}{s - z_2 I}. \tag{20.7}$$

PROOF: We have, by (19.9) and (19.17),

$$D(C(s - z_1 I)^{\gamma_1}(s-z_2 I)^{\gamma_2})$$

$$= C((D(s-z_1 I)^{\gamma_1}(s-z_2 I)^{\gamma_2} + (s-z_1 I)^{\gamma_1}(D(s-z_2 I)^{\gamma_2}))$$

$$= C(\gamma_1(s-z_1 I)^{\gamma_1 - 1}(s-z_2 I)^{\gamma_2} + (s-z_1 I)^{\gamma_1}\gamma_2(s-z_2 I)^{\gamma_2 - 1}).$$

Therefore, (20.6) satisfies

$$\frac{Dy}{y} = \frac{\gamma_1}{s - z_1 I} + \frac{\gamma_2}{s - z_2 I} = \frac{(-2a_2+b_1)s - a_1 + b_0}{a_2 s^2 + a_1 s + a_0 I}.$$

EXAMPLE 20.1. (THE BESSEL DIFFERENTIAL EQUATION). The equation is

$$t^2 z''(t) + tz'(t) + (t^2 - \alpha^2)z(t) = 0, \tag{20.8}$$

where $\alpha$ is a complex number. By the substitution

$$z = t^{-\alpha} y,$$

the above equation becomes

$$ty''(t) - (2\alpha-1)y'(t) + ty(t) = 0. \tag{20.9}$$

Here we have

$$a_2 = 1,\ b_2 = 0,\ a_1 = 0,\ b_1 = 1-2\alpha,\ a_0 = 1,\ b_0 = 0,$$

so $a_2 - b_1 = 2\alpha$. Hence the initial condition (20.3) becomes

$$y(0) = 0 \quad \text{if} \quad \alpha \neq 0. \tag{20.10}$$

The equation (20.4) for (20.9) is

$$\frac{Dy}{y} = \frac{-(2\alpha+1)}{s^2+I} = \frac{-\alpha-1/2}{s+iI} + \frac{-\alpha-1/2}{s-iI}. \tag{20.11}$$

Hence

$$y = C(s+iI)^{-\alpha-1/2}(s-iI)^{-\alpha-1/2} = C(s^2+I)^{-\alpha-1/2}$$

$$= C(h^2(I+h^2)^{-1})^{\alpha+1/2} = C\left(\sum_{k=0}^{\infty} \binom{-\alpha-1/2}{k} h^{2k}\right) h^{2\alpha+1}$$

satisfies (20.11).

We have, in the case $\operatorname{Re}\alpha \geqq 0$,

$$\binom{-\alpha-1/2}{k} = \frac{(-\alpha-1/2)(-\alpha-3/2)\ \dots\ (-\alpha-k+1/2)}{k!}$$

$$= \frac{(2\alpha+1)(2\alpha+3)\ \dots\ (2\alpha+2k-1)(-1)^k}{k!2^k}$$

$$= \frac{(-1)^k}{k!2^k} \frac{(2\alpha+1)(2\alpha+2)\ \dots\ (2\alpha+2k)}{2^k(\alpha+1)\ \dots\ (\alpha+k)}$$

$$= \frac{(-1)^k\Gamma(2\alpha+2k+1)\Gamma(\alpha+1)}{2^{2k}\Gamma(k+1)\Gamma(2\alpha+1)\Gamma(\alpha+k+1)}.$$

Thus we obtain

$$y_\alpha(t) = \sum_{k=0}^{\infty} \frac{(-1)^k}{\Gamma(k+1)\Gamma(\alpha+k+1)} \left(\frac{t}{2}\right)^{2k+2\alpha} \text{ by taking}$$

$$C = \frac{\Gamma(2\alpha+1)}{\Gamma(\alpha+1)2^{2\alpha}}. \tag{20.12}$$

If $\operatorname{Re}\alpha > 1$, then $y_\alpha(t)$ is twice continuously differentiable in $t$ for $t \geqq 0$. Hence the solution $y_\alpha(t)$ of (20.11) satisfying (20.10) is also a solution of (20.9) for $t \geqq 0$. This means that, when $t \geqq 0$ and $\operatorname{Re}\alpha > 1$, the coefficients of $t^{2k+2\alpha-1}$ in the infinite series in $t$ given by

$$ty_\alpha''(t) - (2\alpha-1)y_\alpha'(t) + ty_\alpha(t)$$

must vanish as analytic functions of $\alpha$ $(k = 0,1,2,\dots)$.

Therefore, since $y_\alpha(t)$ with $\operatorname{Re}\alpha \geqq 0$ is twice continuously differentiable in $t > 0$, we see, as in the case of $\operatorname{Re}\alpha > 1$, that the formula

$$ty''_\alpha(t) - (2\alpha-1)y'_\alpha(t) + ty_\alpha(t)$$

must vanish when $t > 0$, because the coefficients of $t^{2k+2\alpha-1}$ all vanish in the above formula.

Thus we have proved that, when $\operatorname{Re}\alpha > 0$ or $\alpha = 0$, $y_\alpha(t)$ given by (20.12) is a solution of (20.9) at every $t > 0$ and satisfies (20.10).

In this way, we have obtained the *Bessel function of the first kind and of order* $\alpha$ ($\operatorname{Re}\alpha > 0$ or $\alpha = 0$)

$$J_\alpha(t) = t^{-\alpha}y_\alpha(t) = \sum_{k=0}^{\infty} \frac{(-1)^k}{\Gamma(k+1)\Gamma(\alpha+k+1)} \left(\frac{t}{2}\right)^{2k+\alpha} \tag{20.13}$$

which satisfies the original Bessel differential equation (20.8) for $t > 0$.

REMARK 20.1. In order to obtain another solution $\hat{y}_\alpha(t)$ of (20.9) which is linearly independent of $y_\alpha(t)$, we shall use the so-called *D'Alembert method*.* We put

$$\hat{y}_\alpha(t) = y_\alpha(t)x(t), \tag{20.14}$$

and from

$$\begin{cases} ty''_\alpha(t) - (2\alpha-1)y'_\alpha(t) + ty_\alpha(t) = 0, \\ t\hat{y}''_\alpha(t) - (2\alpha-1)\hat{y}'_\alpha(t) + t\hat{y}_\alpha(t) = 0, \end{cases}$$

we obtain

$$ty_\alpha(t)x''(t) + (2ty'_\alpha(t) - (2\alpha-1)y_\alpha(t))x'(t) = 0. \tag{20.15}$$

This is a linear ordinary differential equation of the first order for the unknown $x'(t)$. Hence, by knowing $y_\alpha(t) \not\equiv 0$, we can obtain a solution $x'(t) \not\equiv 0$ of (20.15). Therefore, we may take

$$x(t)y_\alpha(t) \quad \text{for} \quad \hat{y}_\alpha(t),$$

where $x(t)$ is not a constant since it is a primitive of $x'(t) \not\equiv 0$. The same remark applies to the following Examples.

EXAMPLE 20.2. (THE CONFLUENT HYPERGEOMETRIC DIFFERENTIAL EQUATION). The equation is

*Jean le Rond D'Alembert (1717-1783).

$$ty''(t) + (c-t)y'(t) - ay(t) = 0, \tag{20.16}$$

where $c$ and $a$ are complex numbers.* Here we have

$$a_2 = 1,\ b_2 = 0,\ a_1 = -1,\ b_1 = c,\ a_0 = 0,\ b_0 = -a,$$

so $(a_2-b_1) = 1 - c$. Thus the initial condition (20.3) becomes

$$y(0) = 0 \quad \text{if} \quad c \neq 1. \tag{20.17}$$

The equation (20.4) for (20.16) is

$$\frac{Dy}{y} = \frac{(-2+c)s + 1 - a}{s^2-s} = \frac{a-1}{s} + \frac{c-a-1}{s-I}. \tag{20.18}$$

Thus, $C_1$ being a non-zero constant,

$$y = C_1 s^{a-1}(s-I)^{c-a-1} = C_1 h^{1-a}\left(\frac{I-h}{h}\right)^{c-a-1}$$

$$= C_1 h^{1-a-c+a+1}(I-h)^{c-a-1}$$

$$= C_1 \sum_{k=0}^{\infty} \binom{c-a-1}{k}(-1)^k h^k h^{2-c}$$

$$= C_1 \sum_{k=0}^{\infty} \binom{c-a-1}{k}(-1)^k \Gamma(k+2-c)^{-1} t^{k+1-c}.$$

When $\mathrm{Re}(1-c) > 0$ or $1-c = 0$, we have

$$\binom{c-a-1}{k}(-1)^k \frac{t^{k+1-c}}{\Gamma(k+2-c)} = \frac{(a-c+1)(a-c+2)\ \ldots\ (a-c+k)}{\Gamma(2-c)(2-c+1)\ \ldots\ (2-c+k-1)} \frac{t^{k+1-c}}{k!}.$$

Thus, as in the case of $y_\alpha(t)$, we have the following result: If $\mathrm{Re}(1-c) > 0$ or $(1-c) = 0$,

then

$$y_{c,a}(t) = t^{1-c} \sum_{k=0}^{\infty} \frac{(a-c+1)(a-c+2)\ \ldots\ (a-c+k)}{\Gamma(2-c)(2-c+1)\ \ldots\ (2-c+k-1)} \frac{t^k}{k!} \tag{20.19}$$

is a solution of (20.16) for every $t > 0$ and satisfies (20.17).

EXAMPLE 20.3. (THE LAGUERRE** DIFFERENTIAL EQUATION). The equation is

---

*This equation is obtained from the *Gauss differential equation*

$$t(1-kt)y''(t) + (c-bt)y'(t) - ay(t) = 0.$$

by *confluence*, i.e., by letting $k \to 0$ and $b \to 1$.

**Edmond Nicolas Laguerre (1834-1886).

$$ty''(t) - (t+\alpha-1)y'(t) + (\alpha+\lambda)y(t) = 0, \qquad (20.20)$$

($\alpha$ and $\lambda$ are complex numbers),

so that it is essentially the same as the confluent hypergeometric differential equation.

For the equation (20.20), we have

$$a_2 = 1, \quad b_2 = 0, \quad a_1 = -1, \quad b_1 = 1-\alpha, \quad a_0 = 0, \quad b_0 = \alpha+\lambda,$$

so $a_2 - b_1 = \alpha$. Hence the initial condition becomes

$$y(0) = 0 \quad \text{if} \quad \alpha \neq 0. \qquad (20.21)$$

The equation (20.4) for (20.20) becomes

$$\frac{Dy}{y} = \frac{(-2+1-\alpha)s + 1 + \alpha + \lambda}{s^2 - s} = \frac{-1-\alpha-\lambda}{s} + \frac{\lambda}{s-I}. \qquad (20.22)$$

Thus, for $\mathrm{Re}\,\alpha > 0$ or $\alpha = 0$,

$$\begin{aligned} y_{\alpha,\lambda} &= Cs^{-1-\alpha-\lambda}(s-I)^{\lambda} = Ch^{1+\alpha}(I-h)^{\lambda} \\ & C\sum_{k=0}^{\infty}\binom{\lambda}{k}(-1)^k\Gamma(k+\alpha+1)^{-1}t^{k+\alpha} \end{aligned} \qquad (20.23)$$

is a solution of (20.22) satisfying (20.21). The proof is the same as in Example 20.1.

Moreover, $t^{-\alpha}y_{\alpha,\lambda}$ reduces to a polynomial in $t$ if and only if $\lambda = 0,1,2,\ldots$ . Furthermore, when

$$\lambda = n \quad \text{and} \quad C = \Gamma(\alpha+n+1)\Gamma(n+1)^{-1},$$

we obtain the *n-th Laguerre polynomial of order* $\alpha$,

$$\begin{aligned} L_n^{(\alpha)}(t) - t^{-\alpha}y_{\alpha,n} &= \frac{\Gamma(n+\alpha+1)}{n!}\sum_{k=0}^{n}\binom{n}{k}\frac{(-t)^k}{\Gamma(k+\alpha+1)} \\ &= \sum_{k=0}^{n}\binom{n+\alpha}{n-k}\frac{(-t)^k}{k!}, \end{aligned} \qquad (20.24)$$

because

$$\begin{aligned} &\binom{n}{k}\frac{\Gamma(\alpha+n+1)}{n!}\frac{1}{\Gamma(\alpha+k+1)} \\ &= \frac{n(n-1)\ldots(n-k+1)}{k!}\;\frac{(\alpha+n)(\alpha+n-1)\ldots(\alpha+k+1)\Gamma(\alpha+k+1)}{n(n-1)\ldots(n-k+1)((n-k)!)\Gamma(\alpha+k+1)} \\ &= \binom{n+\alpha}{n-k}\frac{1}{k!}. \end{aligned}$$

Usually, $L_n(t) = L_n^{(0)}(t)$ is called the *n-th Laguerre polynomial* and we have

$$L_0(t) = 1, \quad L_1(t) = -t+1, \quad L_2(t) = \frac{t^2}{2} - 2t+1, \ldots .$$

A COMPLEMENT TO THEOREM 18. Consider the equation

$$\frac{Dy}{y} = \frac{\gamma}{s-\alpha I} + \frac{\beta}{(s-\alpha I)^2} \quad (\alpha,\ \beta \text{ are complex numbers}), \tag{20.25}$$

in the case that the algebraic equation

$$a_2z^2 + a_1z + a_0 = 0$$

has a double root $\alpha$. In this case, we may obtain a solution $y$ of (20.25):

$$\begin{aligned} &y = C(s - \alpha I)^{\gamma}\hat{y}, \quad \text{where} \\ &\hat{y} = T^{\alpha}e^{-\beta h} \in C/C \quad (\text{but not } \in C.) \end{aligned} \tag{20.26}$$

Here $T^{\alpha}$ is the mapping* of $C/C$ into $C/C$ defined by

$$T^{\alpha}\frac{f}{g} = \frac{T^{\alpha}f}{T^{\alpha}g} = \frac{\{e^{\alpha t}f(t)\}}{\{e^{\alpha t}g(t)\}} \quad (\text{when } f,g \in C \text{ and } g \neq 0). \tag{20.27}$$

PROOF OF THE CONSISTENCY OF (20.27) WITH $T^{\alpha}f = \{e^{\alpha t}f(t)\}$: Starting with $T^{\alpha}f = \{e^{\alpha t}f(t)\}$ we have

$$\begin{aligned} T^{\alpha}(fk) &= \left\{e^{\alpha t}\int_0^t f(t-u)k(u)\,du\right\} \quad (\text{when } f,k \in C) \\ &= \left\{\int_0^t e^{\alpha(t-u)}f(t-u)e^{\alpha u}k(u)\,du\right\} \\ &= (T^{\alpha}f)(T^{\alpha}k), \end{aligned} \tag{20.28}$$

and

$$T^{\alpha}(\beta I) = \frac{T^{\alpha}(\beta h)}{T^{\alpha}h} = \frac{\beta(T^{\alpha}h)}{T^{\alpha}h} = \beta I. \tag{20.29}$$

Thus

$$T^{\alpha}\frac{a}{b} = \frac{T^{\alpha}a}{T^{\alpha}b} \quad \text{for} \quad a = \frac{f}{g}, \quad b = \frac{f_1}{g_1} \tag{20.30}$$
$$(f,g,f_1,g_1 \in C \text{ and } g \neq 0,\ g_1 \neq 0 \text{ and } f_1 \neq 0),$$

---
*Due to J. Mikusiński [5], p. 246ff; cf. §37 of the present book.

because, by (20.28),

$$T^{\alpha}\frac{a}{b} = T^{\alpha}\frac{fg_1}{gf_1} = \frac{T^{\alpha}(fg_1)}{T^{\alpha}(gf_1)} = \frac{(T^{\alpha}f)(T^{\alpha}g_1)}{(T^{\alpha}g)(T^{\alpha}f_1)}$$
$$= \frac{T^{\alpha}f}{T^{\alpha}g} \Big/ \frac{T^{\alpha}f_1}{T^{\alpha}g_1} = \frac{T^{\alpha}a}{T^{\alpha}b}.$$

In this way, we have proved the above consistency.

Furthermore, we have

$$T^{-\alpha}(D\hat{y}) = D(T^{-\alpha}\hat{y}) \quad \text{for } f \text{ and } g \in C \text{ such that } \hat{y} = \frac{f}{g}. \tag{20.31}$$

In fact, we obtain, by (20.27) and (19.12),

$$T^{-\alpha}(D\hat{y}) = T^{-\alpha}(D\frac{f}{g}) = T^{-\alpha}\frac{(Df)g - f(Dg)}{g^2}$$
$$= \frac{T^{-\alpha}((Df)g - f(Dg))}{T^{-\alpha}g^2} = \frac{(T^{-\alpha}(Df))(T^{-\alpha}g) - (T^{-\alpha}f)(T^{-\alpha}(Dg))}{(T^{-\alpha}g)(T^{-\alpha}g)}$$
$$= \frac{(D(T^{-\alpha}f))(T^{-\alpha}g) - (T^{-\alpha}f)(D(T^{-\alpha}g))}{(T^{-\alpha}g)(T^{-\alpha}g)} = D\frac{T^{-\alpha}f}{T^{-\alpha}g} = D(T^{-\alpha}\hat{y}).$$

As a corollary of (20.31), we have

$$\begin{cases} \text{For } \hat{y} = \frac{f}{g} \text{ with } f,g \in C, \\ \frac{D\hat{y}}{\hat{y}} = \frac{\beta I}{(s-\alpha I)^2} \text{ is equivalent to } \frac{DT^{-\alpha}\hat{y}}{T^{-\alpha}\hat{y}} = \frac{\beta I}{s^2}. \end{cases} \tag{20.32}$$

THE PROOF IS EASY. In fact, we have

$$\frac{DT^{-\alpha}\hat{y}}{T^{-\alpha}\hat{y}} = T^{-\alpha}\frac{D\hat{y}}{\hat{y}} = T^{-\alpha}\frac{\beta I}{(s-\alpha I)^2} = (T^{-\alpha}\beta)(T^{-\alpha}\{e^{\alpha t}\})^2$$
$$= \beta I\{e^{-\alpha t}e^{\alpha t}\}^2 = \beta I h^2 = \frac{\beta I}{s^2};$$

and, vice versa,

$$T^{\alpha}T^{-\alpha}\frac{D\hat{y}}{\hat{y}} = \frac{D\hat{y}}{\hat{y}} = T^{\alpha}\frac{\beta I}{s^2} = T^{\alpha}\beta h^2 = (T^{\alpha}\beta I)(T^{\alpha}h)^2$$
$$= \beta I\{e^{\alpha t}\}^2 = \beta I\frac{I}{(s-\alpha I)^2}.$$

Hence, motivated by (19.19), we shall take

$$T^{-\alpha}\hat{y} = z = e^{-\beta/s} = e^{-\beta h} = I + \sum_{k=1}^{\infty} \frac{(-\beta)^k h^k}{k!}, \tag{20.33}$$

because we have, by (19.19),

$$Dz = \frac{d}{ds} e^{-\beta/s} = e^{-\beta/s} \frac{\beta I}{s^2} = \frac{\beta z}{s^2}. \tag{20.34}$$

Therefore, by (20.33), we have obtained the proof of (20.26):

$$\hat{y} = T^{\alpha}T^{-\alpha}\hat{y} = T^{\alpha}e^{-\beta h}. \tag{20.26'}$$

REMARK 20.1. We already know (16.13):

$$e^{-\beta/s} = e^{-\beta h} = s\{J_0(2\sqrt{\beta t})\} = \frac{I}{h}\{J_0(2\sqrt{\beta t})\}, \tag{16.13'}$$

so that, by (20.27),

$$\begin{aligned} \hat{y} = T^{\alpha}e^{-\beta h} &= T^{\alpha}\frac{\{J_0(2\sqrt{\beta t})\}}{h} = \frac{\{e^{\alpha t}J_0(2\sqrt{\beta t})\}}{\{e^{\alpha t}\}} \\ &= (s-\alpha I)\{e^{\alpha t}J_0(2\sqrt{\beta t})\}. \end{aligned} \tag{20.35}$$

Hence, by (15.3), we see that

$$\hat{y} \in C_H \subset C/C, \quad \text{but} \quad \hat{y} \notin C \quad \text{since} \quad [J_0(0)] \neq 0.$$

EXAMPLE 20.4. For the equation

$$\alpha t y''(t) + y(t) = 0, \tag{20.36}$$

we have

$$a_2 = \alpha, \quad b_2 = 0, \quad a_1 = 0, \quad b_1 = 0, \quad a_0 = 0, \quad b_0 = 1,$$

so $a_2 - b_1 = \alpha$. Hence the initial condition (20.3) for (20.36) becomes

$$y(0) = 0 \quad \text{if} \quad \alpha \neq 0. \tag{20.37}$$

Furthermore, (20.4) for (20.36) is given by

$$\frac{Dy}{y} = \frac{-2\alpha s + 1}{s^2} = \frac{-2\alpha}{s} + \frac{I}{s^2}. \tag{20.38}$$

Hence, by (20.35),

$$\begin{aligned} y = Cs^{-2\alpha}e^{-h} &= Ch^{2\alpha} + C\sum_{k=1}^{\infty}\frac{(-1)^k h^{k+2\alpha}}{k!} \\ &= C\left\{\sum_{k=0}^{\infty}\Gamma(k+2\alpha)^{-1}\Gamma(k+1)^{-1}(-1)^k t^{k+2\alpha-1}\right\} \end{aligned} \tag{20.39}$$

is a solution of (20.38) satisfying (20.37) when $\mathrm{Re}(2\alpha-1) > 0$.

If $\mathrm{Re}\,\alpha > \frac{3}{2}$, then (20.39) is twice continuously differentiable in $t$ for $t \geqq 0$. Hence the solution (20.39) of (20.37) - (20.38) is, for $\mathrm{Re}\,\alpha > \frac{3}{2}$, a solution of (20.36) at every $t \geqq 0$.

Therefore, as in Example 20.1, we see that, for $t > 0$ and $\mathrm{Re}\,\alpha > \frac{1}{2}$, (20.39) is a solution of (20.36) satisfying (20.37).

EXERCISES FOR §20.

(α) Verify

$$De^{-\beta/s^m} = De^{-\beta h^m} = e^{-\beta h^m}\frac{m\beta}{s^{m+1}} \qquad (m = 1,2,\dots).$$

(β) Solve ($b$ and $c$ are real numbers)

$$ty''(t) + 2(b+1)y'(t) + (c + \tfrac{t}{4})y(t) = 0.$$

(γ) For complex numbers $a_j$'s and $b_j$'s, show that the equation

$$a_3ty'''(t) + (a_2t+b_2)y''(t) + (a_1t+b_1)y'(t) + (a_0t+b_0)y(t) = 0$$

converts into

$$\begin{aligned} &Dy(-a_3s^3 - a_2s^2 - a_1s - a_0) \\ &\quad + y((-3a_3+b_2)s^2 + (-2a_2+b_1)s + (-a_1+b_0)) \\ &\quad + [y'(0)](a_3-b_2) + [y(0)]((2a_3-b_2)s + (a_2-b_1)) = 0. \end{aligned}$$

(δ) Under the initial conditions

$$\begin{cases} y(0) = 0 \text{ if } ((2a_3-b_2)s + (a_2-b_1)) \neq 0, \\ y'(0) = 0 \text{ if } a_3-b_2 \neq 0, \end{cases}$$

show that the third order differential equation in (γ) converts into

$$\frac{Dy}{y} = \frac{(-3a_3+b_2)s^2 + (-2a_2+b_1)s - a_1 + b_0}{a_3s^3 + a_2s^2 + a_1s + a_0}.$$

REMARK 20.2. The method of integration explained above in this section can be extended to the integration of the n-th order linear differential equation with linear coefficients:

$$a_nty^{(n)}(t) + (a_{n-1}t+b_{n-1})y^{(n-1)}(t) + \dots + (a_0t+b_0)y(t) = 0.$$

## §21. SUPPLEMENTS. I: WEIERSTRASS' POLYNOMIAL APPROXIMATION THEOREM. II: MIKUSIŃSKI'S THEOREM OF MOMENTS

I. The following formulation and proof of the theorem of Weierstrass is due to Bernstein.*

THEOREM 19. Let $f(t)$ be a complex-valued continuous function defined on the closed interval $[0,1]$. Then, defining polynomials

$$P_n(t) = \sum_{p=0}^{n} {}_nC_p f(\tfrac{p}{n}) t^p (1-t)^{n-p}, \tag{21.1}$$

we have

$$\lim_{n\to\infty} \max_{0\leq t\leq 1} |f(t) - P_n(t)| = 0. \tag{21.2}$$

PROOF: Differentiating

$$(t+u)^n = \sum_{p=0}^{n} {}_nC_p t^p u^{n-p} \tag{21.3}$$

with respect to $t$ and multiplying by $t$, we obtain

$$nt(t+u)^{n-1} = \sum_{p=0}^{n} p\,{}_nC_p t^p u^{n-p}. \tag{21.4}$$

Similarly, differentiating (2.13) twice with respect to $t$ and multiplying by $t^2$, we obtain

$$n(n-1)t^2(t+u)^{n-2} = \sum_{p=0}^{n} p(p-1)\,{}_nC_p t^p u^{n-p}. \tag{21.5}$$

Thus, if we put

$$r_p(t) = {}_nC_p t^p (1-t)^{n-p}, \tag{21.6}$$

we have

$$\begin{cases} \sum_{p=0}^{n} r_p(t) = 1, \quad \sum_{p=0}^{n} p\cdot r_p(t) = nt \\ \sum_{p=0}^{n} p(p-1) r_p(t) = n(n-1)t^2. \end{cases} \tag{21.7}$$

Hence

---
*Sergei Natanovic Bernstein (1880-1968).

$$\sum_{p=0}^{n}(p-nt)^2 r_p(t)$$

$$= n^2t^2 \sum_{p=0}^{n} r_p(t) - 2nt \sum_{p=0}^{n} pr_p(t) + \sum_{p=0}^{n} p^2 r_p(t)$$

$$= n^2t^2 - 2nt.nt + (nt + n(n-1)t^2) = nt(1-t),$$

i.e.,

$$\sum_{p=0}^{n} (p-nt)^2 r_p(t) = nt(1 - t). \tag{21.8}$$

By the uniform continuity of the function $f(t)$, there exists, for any $\varepsilon > 0$, a $\delta > 0$ such that

$$|f(t) - f(t')| < \varepsilon \text{ whenever } |t - t'| < \delta. \tag{21.9}$$

Also by the continuity of $f(t)$, we have

$$\max_{0 \leqq t \leqq 1} |f(t)| = M < +\infty. \tag{21.10}$$

Therefore, by (21.7), we obtain

$$\left|f(t) - \sum_{p=0}^{n} f(\tfrac{p}{n}) r_p(t)\right| = \left|\sum_{p=0}^{n} (f(t) - f(\tfrac{p}{n})) r_p(t)\right|$$

$$\leqq \left|\sum_{|p-nt| \leqq \delta n}\right| + \left|\sum_{|p-nt| > \delta n}\right|.$$

We have, by $r_p(t) \geqq 0$, (21.7) and (21.9),

$$\left|\sum_{|p-nt| \leqq \delta n}\right| \leqq \varepsilon \sum_{p=0}^{n} r_p(t) = \varepsilon.$$

We also have, by (21.8) and (21.10),

$$\left|\sum_{|p-nt| > \delta n}\right| \leqq 2M \sum_{|p-nt| > \delta n} r_p(t)$$

$$\leqq \frac{2M}{n^2\delta^2} \sum_{p=0}^{n} (p-nt)^2 r_p(t) \leqq \frac{2Mt(1-t)}{n\delta^2} \leqq \frac{M}{2\delta^2 n}.$$

The extreme right term tends to $0$ as $n \to \infty$, and thus

$$\lim_{n\to\infty} \max_{0 \leqq t \leqq 1} |f(t) - P_n(t)| \leqq \varepsilon.$$

Since $\varepsilon > 0$ was arbitrarily chosen, we have proved (21.2).

We shall now state and prove II.

THEOREM 20. Let a complex-valued continuous function $w(t)$ defined on the closed interval $[0,T]$, $0 < T < \infty$, satisfy

$$\left|\int_0^T e^{nt} w(t)\,dt\right| \leqq M < \infty \qquad (n = 1,2,3,\ldots), \tag{21.11}$$

where $M$ is a constant. Then $w(t) \equiv 0$ on $[0,T]$.

PROOF: Let $k$ and $x$ be natural numbers, and put $w(T-u) = g(u)$. Then, by (21.11), we have

$$\left|\int_0^T e^{kx(T-t)} g(t)\,dt\right| = \left|\int_0^T e^{kx(T-u)} w(T-u)\,du\right| \leqq M$$

$$(k = 1,2,3,\ldots;\quad x = 1,2,3,\ldots).$$

Hence

$$\left|\sum_{k=1}^{\infty} \frac{(-1)^{k-1}}{k!} e^{-kx(T-t)} \int_0^T e^{kx(T-u)} g(u)\,du\right|$$
$$\leqq \sum_{k=1}^{\infty} \frac{1}{k!} e^{-kx(T-t)} M \leqq M(\exp(e^{-x(T-t)}) - 1),$$

and so, for any $\varepsilon > 0$ with $\varepsilon < T$, the extreme right term tends to $0$ as $x \to \infty$ uniformly in $t$ on $0 \leqq t \leqq T-\varepsilon$. Thus

$$\begin{cases} \lim\limits_{x\to\infty} \sum\limits_{k=1}^{\infty} \dfrac{(-1)^{k-1}}{k!} e^{-kx(T-t)} \displaystyle\int_0^T e^{kx(T-u)} g(u)\,du = 0 \\ \text{uniformly in } t \quad (0 \leqq t \leqq T-\varepsilon). \end{cases} \tag{21.12}$$

Since we have

$$\lim_{n\to\infty} \sum_{k=n}^{\infty} \frac{1}{k!} = 0$$

and

$$0 \leqq e^{-kx(T-t)} \leqq 1 \qquad (0 \leqq t \leqq T;\quad k = 1,2,\ldots),$$

we obtain

$$\sum_{k=1}^{\infty} \frac{(-1)^{k-1}}{k!} e^{-kx(T-t)} \int_0^T e^{kx(T-u)} g(u)\,du$$
$$= \int_0^T \sum_{k=1}^{\infty} \frac{(-1)^{k-1}}{k!} e^{kx(t-u)} g(u)\,du$$
$$= \int_0^T (1 - \exp(-e^{x(t-u)}))g(u)\,du.$$

This implies, by (21.12),

$$\begin{cases} \lim_{x\to\infty} \int_0^T (1 - \exp(-e^{-x(t-u)}) g(u)\,du = 0 \\ \text{uniformly in } t \quad (0 \leqq t \leqq T - \varepsilon). \end{cases} \tag{21.13}$$

Since the integrand

$$(1 - \exp(-e^{x(t-u)}))g(u)$$

is continuous and bounded on the domain defined by

$$0 \leqq t \leqq T - \varepsilon, \quad 0 \leqq u \leqq T, \quad 0 < x,$$

we can exchange the order of the limit and the integral in (21.13). Hence

$$0 = \int_0^T \lim_{x\to\infty} (1 - \exp(-e^{x(t-u)}))g(u)\,du$$
$$= \int_0^T \begin{Bmatrix} 1, & t > u \\ 0, & t < u \end{Bmatrix} g(u)\,du = \int_0^t g(u)\,du \qquad (0 \leqq t \leqq T - \varepsilon).$$

We have thus obtained

$$\int_0^t g(u)\,du = 0 \qquad (0 \leqq t \leqq T),$$

because of the arbitrary choice of $\varepsilon$ $(0 < \varepsilon < T)$. This proves that the continuous function $g(t)$ vanishes for all $t \in [0,T]$, and so $w(t)$ also vanishes for all $t \in [0,T]$.

REMARK 21.1. The above proof is adapted from J. Mikusiński [5], p. 18. J. Mikusiński and R. Nardzewski applied Theorem 22 in an elementary proof of the Titchmarsh convolution theorem.

# Part III
# Shift Operator $\exp(-\lambda s)$ and Diffusion Operator $\exp(-\lambda s^{1/2})$

## Chapter VIII
## Exponential Hyperfunctions $\exp(-\lambda s)$ and $\exp(-\lambda s^{1/2})$

### §22. SHIFT OPERATOR $\exp(-\lambda s) = e^{-\lambda s}$. FUNCTION SPACE $K = K[0,\infty)$

Let $\lambda \geqq 0$. We call the function $H_\lambda(t)$ defined by

$$H_\lambda(t) = \begin{cases} 0, & 0 \leqq t < \lambda \\ 1, & 0 \leqq \lambda < t < \infty. \end{cases} \tag{22.1}$$

*Heaviside's unit function.** We then define

$$h_1(\lambda,t) = \int_0^t H_\lambda(u)du = \begin{cases} 0, & 0 \leqq t < \lambda \\ t-\lambda, & 0 \leqq \lambda \leqq t < \infty, \end{cases} \tag{22.2}$$

a continuous function of $t \geqq 0$, containing a parameter $\lambda \geqq 0$.

We shall define**

$$e^{-\lambda s} = s^2\{h_1(\lambda,t)\} \tag{22.3}$$

and give

PROPOSITION 17. For $f \in C$, we have

$$e^{-\lambda s}\{f(t)\} = s\begin{Bmatrix} 0, & 0 \leqq t < \lambda \\ \int_0^{t-\lambda} f(u)du, & 0 \leqq \lambda \leqq t < \infty \end{Bmatrix} \tag{22.4}$$

PROOF: We have, by (22.2) and Theorem 3,

---

*We may give any finite value for $H_\lambda(\lambda)$.

**The reason why for the notation $e^{-\lambda s}$ will be explained in §24. (22.3) was first introduced by J. Mikusiński in the form $h^\lambda = s^2\{h_1(\lambda,t)\}$. See his book [5], p. 181.

$$s^2\left\{\int_0^t h_1(\lambda,t-u)f(u)du\right\}$$
$$= s\cdot s\begin{Bmatrix} 0, & 0 \leqq t < \lambda \\ \int_0^{t-\lambda} (t-u-\lambda)f(u)du, & \lambda \leqq t < \infty \end{Bmatrix}$$
$$= s\begin{Bmatrix} 0, & 0 \leqq t < \lambda \\ \int_0^{t-\lambda} f(u)du, & \lambda \leqq t < \infty \end{Bmatrix},$$

because we have, for $\lambda < t$,

$$s\int_0^{t-\lambda} (t-u-\lambda)f(u)du = \frac{d}{dt}\int_0^{t-\lambda} (t-u-\lambda)f(u)du$$
$$+0 = (t-u-\lambda)f(u)\Big|_{u=t-\lambda} + \int_0^{t-\lambda} \frac{d}{dt}(t-u-\lambda)f(u)du.$$

REMARK 22.1. Since $h_1(0,t) = \{t\}$, we have

$$e^{-\lambda s}\Big|_{\lambda=0} = s^2\{t\} = s\{1\} = I \tag{22.5}$$

so that $e^{-os} = I$.

PROPOSITION 17'. For $\lambda > 0$ and $\mu > 0$, we have

$$e^{-\lambda s}e^{-\mu s} = e^{-(\lambda+\mu)s}. \tag{22.6}$$

PROOF: We have to prove

$$s^2\{h_1(\lambda,t)\}s^2\{h_1(\mu,t)\} = s^2\{h_1(\lambda+\mu,t)\}.$$

Since the ring $C/C$ is commutative, we have, similarly as in (22.4),

$$s^2\{h_1(\lambda,t)\}s^2\{h_1(\mu,t)\}$$
$$= s^2s^2\left\{\int_0^t h_1(\lambda,t-u)h_1(\mu,u)du\right\} = s^2s^2\left\{\int_\mu^{t-\lambda} (t-u-\lambda)(u-\mu)du\right\}$$
$$= s^2s\left\{(t-u-\lambda)(u-\mu)\Big|_{u=t-\lambda} + \int_\mu^{t-\lambda} (u-\mu)du\right\}$$
$$= s^2s\left\{\int_\mu^{t-\lambda} (u-\mu)du = s^2\{t-\lambda-\mu\}\right\} \text{ when } 0 \leqq \lambda+\mu < t,$$

and

$$s^2s^2\left\{\int_\mu^{t-\lambda} (t-u-\lambda)(u-\mu)du\right\} = 0 \quad \text{when} \quad 0 \leqq t < \lambda+\mu,$$

because

$$h_1(\lambda,t-u)h_1(\mu,u) > 0 \quad \text{only if both} \quad 0 \leqq \lambda < t-u \quad \text{and} \quad 0 \leq \mu < u$$

are satisfied.

REMARK 22.2. By (22.4) and by the operation of differentiation $s$, we see that the operation of $e^{-\lambda s}$ may be interpreted as follows:

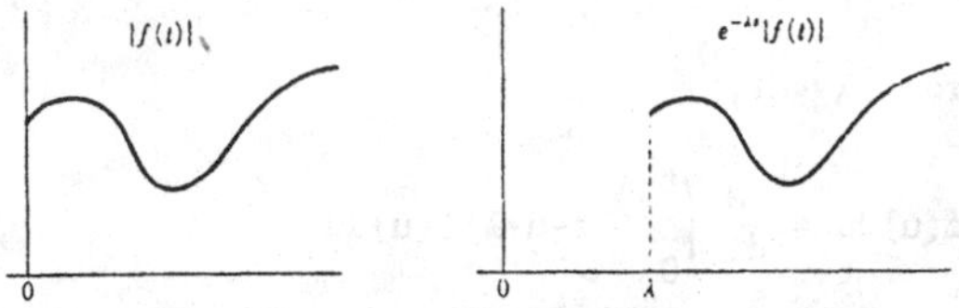


That is, the operator $e^{-\lambda s}$ gives a *shift or translation of the graph of* $\{f(t)\}$ *to the right by length* $\lambda$. We shall thus represent $e^{-\lambda s}\{f(t)\}$ by the function

$$\begin{cases} 0, & 0 \leqq t < \lambda \\ f(t-\lambda), & 0 \leqq \lambda < t \end{cases}$$

which might be discontinuous at $t = \lambda$.

In order to introduce the above mentioned representation, we shall consider the class $K$ of functions.*

THE CLASS $K$ OF FUNCTIONS. Let $K = K[0,\infty)$ be the totality of complex-valued functions $f(t)$ defined on $[0,\infty)$ and satisfying the following conditions:

In every finite interval in $[0,\infty)$, $\{f(t)\}$ has only a finite number of points at which $\{f(t)\}$ is discontinuous; (22.7)

$$\text{For every } t > 0, \quad \int_0^t |f(u)|\,du < \infty. \tag{22.8}$$

Thus, if $f \in K$, the convolution product

$$hf = \{1\}\{f(t)\} = \left\{\int_0^t f(u)\,du\right\}$$

---

*J. Mikusiński [5], p. 106.

exists and $hf \in C$. Moreover, for $f$ and $g \in K$, the convolution

$$(fg)(t) = \left\{\int_0^t f(t-u)g(u)du\right\}$$

exists and $fg \in K$. As in the case of $C$, $K$ constitutes a *commutative ring*. Of course, like $H_\lambda(t)$, a function in $K$ may have a *jump* at its point of discontinuity. Thus for two functions $f$ and $g$, *we regard* $f$ *to be equal to* $g$ *as functions of* $K$ if $f(t) - g(t)$ vanishes at every point $t_0$ at which both $f$ and $g$ are continuous. Moreover, for $f$ and $g \in K$, we define the sum of $f$ and $g$ through

$$(f + g)(t) = f(t) + g(t)$$

at every point $t_0$ where both $f$ and $g$ are continuous. Finally, for $f \in K$ and $\{g(t)\} = \{\int_0^t f(u)du\}$, we have

$$sg = g' + sg(0) = \{f(t)\},$$

the equality in this equation meaning equality at every point $t_0$ where $\{f(t)\}$ is continuous.

EXAMPLE 22.1. For $\{h_1(\lambda,t)\}$ of (22.2), we have

$$s\{h_1(\lambda,t)\} = \begin{Bmatrix} 0, & 0 \leqq t < \lambda \\ 1, & \lambda \leqq t < \infty \end{Bmatrix} = \{H_\lambda(t)\}. \tag{22.9}$$

REMARK 22.3. By (22.3), we can rewrite (22.9) as

$$he^{-\lambda s} = \frac{I}{s} e^{-\lambda s} = \frac{I}{s} s^2\{h_1(\lambda,t)\} = s\{h_1(\lambda,t)\} = \{H_\lambda(t)\}. \qquad (22.9)'$$

EXERCISES FOR §22. Verify the following equalities when $0 \leqq \lambda < \mu$:

(i) $s^{-1}(e^{-\lambda s} - e^{-\mu s}) = \begin{Bmatrix} 1, & \lambda < t < \mu \\ 0, & \text{otherwise} \end{Bmatrix}$.

(ii) $s^{-3}(e^{-\lambda s} - e^{-\mu s}) = \begin{Bmatrix} 0, & 0 \leqq t \leqq \lambda \\ 2^{-1}(t-\lambda)^2, & \lambda \leqq t \leqq \mu \\ 2^{-1}(\mu-\lambda)(2t-\lambda-\mu), & \mu \leqq t \end{Bmatrix}$.

REMARK 22.4. We shall identify $f = \{f(t)\} \in K$ with $\frac{hf}{h} \in C_H$ so that the function $\{f(t)\} \in K$ may be regarded as a concrete representation of the hyperfunction $\frac{hf}{h} \in C_H$.

AN APPLICATION OF $K$ TO PHYSICS. The so-called *impulsive force* in mechanics can be given the following representation as a hyperfunction.

Let $F$ and $\varepsilon$ be positive numbers. If an external force of strength

$\frac{F}{\varepsilon}$ is applied to a particle with unit mass during the time interval between $t = 0$ and $t = \varepsilon$, then we say that, during this time interval, the impulse

$$\frac{F}{\varepsilon} \times \varepsilon = F = \int_0^\varepsilon \frac{F}{\varepsilon}\,du$$

is applied to the particle. We notice that the impulse is equal to the convolution

$$h * f_\varepsilon, \quad \text{where} \quad f_\varepsilon(t) = \left\{\begin{matrix} \frac{F}{\varepsilon}, & 0 \leq t \leq \varepsilon \\ 0, & \varepsilon < t \end{matrix}\right\} \in K.$$

The so-called *impulsive force* is interpreted as a limit of the hyperfunction $f_\varepsilon$ when $\varepsilon$ tends to $0$. Thus

the impulsive force $[F] = [F]I = s\{F\}$,

because, for any $g \in C$,

$$g*f_\varepsilon = \left\{\int_0^\varepsilon g(t-u)\frac{F}{\varepsilon}\,du\right\} = \left\{\frac{F}{\varepsilon}\int_{t-\varepsilon}^t g(\tau)d\tau\right\} \qquad (0 < \varepsilon < t)$$

so that

$$\lim_{\varepsilon \downarrow 0} g*f_\varepsilon = \{Fg(t)\} = [F] \text{ times } \{g(t)\}.$$

Thus, as $\varepsilon \downarrow 0$, the hyperfunction $f_\varepsilon$ tends to the hyperfunction $[F]I$. Physicists call $I$ the *delta function* of Paul Adrian Maurice Dirac (1902- ).

AN EXAMPLE OF THE IMPULSIVE FORCE. Consider the equation

$$y''(t) + 2ay'(t) + b^2y(t) = f; \quad y(0) = 0, \quad y'(0) = 0,$$

where the external force $f$ is given by a hyperfunction $[F]I$. As in the Example 10.6, we have

$$s^2y + 2asy + b^2y = [F]I.$$

Hence

$$\begin{aligned} y &= \frac{[F]I}{s^2+2as+b^2} = \frac{[F]I}{(s+a-i\omega)(s+a+i\omega)} \\ &= \frac{[F]}{2i\omega}\left(\frac{I}{s+a-i\omega} - \frac{I}{s+a+i\omega}\right) \\ &= \left\{\frac{F}{\omega}\left(e^{-at}\,\frac{e^{+i\omega t} - e^{-i\omega t}}{2i}\right)\right\} \\ &= \left\{\frac{F}{\omega}\,e^{-at}\sin\omega t\right\}. \end{aligned}$$

## §23. HYPERFUNCTION-VALUED FUNCTION $f(\lambda)$ AND GENERALIZED DERIVATIVE $\frac{d}{d\lambda} f(\lambda) = f'(\lambda)$.

OPERATOR-VALUED FUNCTION OR HYPERFUNCTION-VALUED FUNCTION. Let, for each fixed value $\lambda_0$ of the parameter $\lambda$, there be given an operator (= hyperfunction) $f(\lambda_0) \in C/C$. Then we shall call $f(\lambda)$ an operator-valued function or a hyperfunction-valued function.

The shift operator

$$s^2\{h_1(\lambda,t)\} \in C_H \qquad (\lambda > 0)$$

is such an example, defined for $\lambda \geqq 0$ if we define

$$\{h_1(0,t)\} = \{t\}.$$

THE GENERALIZED DERIVATIVE.* $f'(\lambda) = \frac{d}{d\lambda} f(\lambda)$ of an operator-valued function $f(\lambda)$ is defined as follows:

Let, for a certain $\frac{b}{a} \neq 0$ of $C/C$,

$$\frac{b}{a} f(\lambda) \text{ be equal to } \{\hat{f}(\lambda,t)\}$$

where $\{\hat{f}(\lambda,t)\}$ is a complex-valued function of $\{\lambda,t\}$ in such a way that, on the domain

$$\mathcal{D} = \{\{\lambda,t\};\ \lambda_1 < \lambda < \lambda_2,\ 0 \leqq t\},$$

both $\hat{f}(\lambda,t)$ and $\frac{\partial}{\partial\lambda} \hat{f}(\lambda,t)$ are continuous functions. Then we say that the *generalized derivative*

$$f'(\lambda) = \frac{d}{d\lambda} f(\lambda)$$

exists for $\lambda_1 < \lambda < \lambda_2$ and is represented by

$$f'(\lambda) = \frac{d}{d\lambda} f(\lambda) = \frac{a}{b} \{\frac{\partial}{\partial\lambda} \hat{f}(\lambda,t)\}. \tag{23.1}$$

REMARK 23.1. The above definition of the generalized derivative is reasonable, i.e., without contradiction. We have to prove the following: If we assume that we have another couple

$$\frac{d}{c} \neq 0 \quad \text{of} \quad C/C \quad \text{and} \quad \hat{\hat{f}}(\lambda,t)$$

for which

---

*J. Mikusiński [5], p. 183.

$$\begin{cases} \text{both } \frac{d}{c} f(\lambda) = \{\hat{\hat{f}}(\lambda,t)\} \text{ and } \{\frac{\partial}{\partial\lambda}\hat{\hat{f}}(\lambda,t)\} \text{ are} \\ \text{continuous functions of } \{\lambda,t\} \text{ on } \mathcal{D}, \end{cases}$$

then we can prove that

$$\frac{a}{b}\{\frac{\partial}{\partial\lambda}\hat{f}(\lambda,t)\} = \frac{c}{d}\{\frac{\partial}{\partial\lambda}\hat{\hat{f}}(\lambda,t)\}. \tag{23.2}$$

PROOF: Multiplying the equality

$$\frac{a}{b}\{\hat{f}(\lambda,t)\} = \frac{c}{d}\{\hat{\hat{f}}(\lambda,t)\} = f(\lambda)$$

by $bd \in C$, we obtain

$$ad\{\hat{f}(\lambda,t)\} = bc\{\hat{\hat{f}}(\lambda,t)\} = bdf(\lambda);$$

that is,

$$\int_0^t ad(t-u)\hat{f}(\lambda,u)du = \int_0^t bc(t-u)\hat{\hat{f}}(\lambda,u)du.$$

Differentiating with respect to $\lambda$, we have

$$\int_0^t ad(t-u)\frac{\partial}{\partial\lambda}\hat{f}(\lambda,u)du = \int_0^t bc(t-u)\frac{\partial}{\partial\lambda}\hat{\hat{f}}(\lambda,u)du$$

i.e.,

$$ad\{\frac{\partial}{\partial\lambda}\hat{f}(\lambda,t)\} = bc\{\frac{\partial}{\partial\lambda}\hat{\hat{f}}(\lambda,t)\}.$$

Multiplying by $\frac{I}{bd}$, we obtain (23.2).

As an application of the above definition, we prove

PROPOSITION 18. For $f(\lambda) = s^2\{h_1(\lambda,t)\}$, we obtain

$$f'(\lambda) = -sf(\lambda) \quad (\lambda > 0), \quad f(0) = I. \tag{23.3}$$

PROOF: We have

$$h^3 f(\lambda) = h\{h_1(\lambda,t)\}$$

$$= h\begin{Bmatrix} 0, & 0 \leqq t < \lambda \\ t-\lambda, & 0 < \lambda \leqq t \end{Bmatrix} = \begin{Bmatrix} 0, & 0 \leqq t < \lambda \\ (t-\lambda)^2/2, & 0 < \lambda \leqq t \end{Bmatrix} = \{\hat{f}(\lambda,t)\}.$$

Hence

$$f'(\lambda) = \frac{I}{h^3}\{\frac{\partial}{\partial\lambda}\hat{f}(\lambda,t)\} = s^3\{-h_1(\lambda,t)\}$$

$$= -s\cdot s^2\{h_1(\lambda,t)\} = -sf(\lambda)$$

and

$$f(0) = s^2\{h_1(0,t)\} = s^2\{t\} = s\{1\} = I$$

REMARK 23.2. In the above, $\{\hat{f}(\lambda,t)\}$ is defined and continuous on $\{\{\lambda,t\};\ 0 \leqq \lambda,\ 0 \leqq t\}$ so that

$$f(\lambda) = s^3\{\hat{f}(\lambda,t)\}$$

is *continuous with respect to* $\lambda$ *in generalized sense*. This means that, by multiplying $f(\lambda)$ by $h^3 \neq 0$ of $C/C$, we obtain $\{\hat{f}(\lambda,t)\}$ which is continuous in $\lambda$. Thus, for any fixed $\lambda_0 \geqq 0$,

$$\begin{aligned} \text{generalized} \lim_{\lambda\to\lambda_0} f(\lambda)^* &= \frac{I}{h^3}\{\lim_{\lambda\to\lambda_0} h^3 f(\lambda)\} \\ &= s^3\{\lim_{\lambda\to\lambda_0} \hat{f}(\lambda,t)\} = s^3\{\hat{f}(\lambda_0,t)\} = f(\lambda_0). \end{aligned} \tag{23.4}$$

Hence $f(\lambda)$ is *continuous at every* $\lambda_0 \geqq 0$ *in the generalized sense*.

We next give a useful proposition regarding the *generalized derivative*.**

PROPOSITION 19. (i). If $f(\lambda)$ has a generalized derivative for $\lambda \in (\lambda_1,\lambda_2)$, then, for any $\frac{q}{p} \neq 0$ of $C/C$, $\frac{q}{p} f(\lambda)$ has the generalized derivative for $\lambda \in (\lambda_1,\lambda_2)$ given by

$$\frac{d}{d\lambda}(\frac{q}{p} f(\lambda)) = \frac{q}{p}\frac{df(\lambda)}{d\lambda} = \frac{q}{p} f'(\lambda). \tag{23.5}$$

(ii). If both $f(\lambda)$ and $g(\lambda)$ have generalized derivatives for $\lambda \in (\lambda_1,\lambda_2)$, then $f(\lambda) \pm g(\lambda)$ also have generalized derivatives for $\lambda \in (\lambda_1,\lambda_2)$ and

$$(f(\lambda) \pm g(\lambda))' = f'(\lambda) \pm g'(\lambda). \tag{23.6}$$

(iii). If both $f(\lambda)$ and $g(\lambda)$ have generalized derivatives for $\lambda \in (\lambda_1,\lambda_2)$, then $f(\lambda)g(\lambda)$ also has a generalized derivative and

$$[f(\lambda)g(\lambda)]' = f'(\lambda)g(\lambda) + f(\lambda)g'(\lambda). \tag{23.7}$$

---

*Let, for a certain $\frac{b}{a} \neq 0$ of $C/C$, $\frac{b}{a} f(\lambda)$ be equal to $\{\hat{f}(\lambda,t)\}$ in such a way that $\lim_{\lambda\to\lambda_0} \hat{f}(\lambda,t)$ exists and is in $C[0,\infty)$ as a function of $t$. Then $\frac{a}{b}\{\lim_{\lambda\to\lambda_0} \hat{f}(\lambda,t)\}$ is, by definition, the *generalized limit* of $f(\lambda)$ as $\lambda \to \lambda_0$. That the value of the generalized limit of $f(\lambda)$ as $\lambda \to \lambda_0$ is independent of the choice of $\frac{b}{a}$ shall be proved similarly as the case of the generalized derivative; see §25.

**J. Mikusiński [5], p. 185-.

(iv). $f(\lambda)$ has the generalized derivative $f'(\lambda) = 0$ for $\lambda \in (\lambda_1,\lambda_2)$ if and only if

$$f(\lambda) \text{ is equal to a fixed hyperfunction } c \in C/C. \tag{23.8}$$

PROOF: (i) Let both $\{\hat{f}(\lambda,t)\} = \frac{b}{a} f(\lambda)$ $(\frac{b}{a} \neq 0)$ and $\{\frac{\partial}{\partial\lambda} \hat{f}(\lambda,t)\}$ be continuous on $\mathcal{D} = \{\{\lambda,t\};\ \lambda_1 < \lambda < \lambda_2,\ 0 \leqq t\}$:

$$f'(\lambda) = \frac{a}{b}\{\frac{\partial}{\partial\lambda} \hat{f}(\lambda,t)\}.$$

Then, by

$$\frac{p}{q}\frac{b}{a}(\frac{q}{p} f(\lambda)) = \frac{b}{a} f(\lambda) = \{\hat{f}(\lambda,t)\},$$

we see that $\frac{q}{p} f(\lambda)$ has a generalized derivative and

$$(\frac{q}{p} f(\lambda))' = \frac{qa}{pb} \{\frac{\partial}{\partial\lambda} \hat{f}(\lambda,t)\} = \frac{q}{p}(\frac{a}{b} \{\frac{\partial}{\partial\lambda} \hat{f}(\lambda,t)\}) = \frac{q}{p} f'(\lambda).$$

(ii). Let $\{\hat{f}(\lambda,t)\} = \frac{b}{a} f(\lambda)$ with $\frac{b}{a} \neq 0$ and $\{g(\lambda,t)\} = \frac{d}{c} g(\lambda)$ with $\frac{d}{c} \neq 0$ such that

$$f'(\lambda) = \frac{a}{b} \{\frac{\partial}{\partial\lambda} \hat{f}(\lambda,t)\} \quad \text{and} \quad g'(\lambda) = \frac{c}{d} \{\frac{\partial}{\partial\lambda} \hat{g}(\lambda,t)\}.$$

Then we have

$$f(\lambda) = \frac{a}{b}\{\hat{f}(\lambda,t)\} = \frac{I}{bd} \{\check{f}(\lambda,t)\}$$

$$\text{with } \check{f}(\lambda,t) = \int_0^t (ad)(t-u)\hat{f}(\lambda,u)du \quad \text{and} \quad f'(\lambda) = \frac{I}{bd} \{\frac{\partial}{\partial\lambda} \check{f}(\lambda,t)\},$$

and

$$g(\lambda) = \frac{c}{d}\{\hat{g}(\lambda,t)\} = \frac{I}{bd} \{\check{g}(\lambda,t)\}$$

$$\text{with } \check{g}(\lambda,t) = \int_0^t (bc)(t-u)\hat{g}(\lambda,u)du \quad \text{and} \quad g'(\lambda) = \frac{I}{bd} \{\frac{\partial}{\partial\lambda} \check{g}(\lambda,t)\}.$$

Therefore

$$f(\lambda) \pm g(\lambda) = \frac{I}{bd} \{\check{f}(\lambda,t) \pm \check{g}(\lambda,t)\}$$

and

$$(f(\lambda) \pm g(\lambda))' = \frac{I}{bd} \{\frac{\partial}{\partial\lambda} \check{f}(\lambda,t) \pm \frac{\partial}{\partial\lambda} \check{g}(\lambda,t)\} = f'(\lambda) + g'(\lambda).$$

(iii). Let, as above,

$$f(\lambda) = \frac{I}{bd} \{\check{f}(\lambda,t)\} \quad \text{and} \quad f'(\lambda) = \frac{I}{bd} \{\frac{\partial}{\partial\lambda} \check{f}(\lambda,t)\}$$

and

$$g(\lambda) = \frac{I}{bd}\{\check{g}(\lambda,t)\} \quad \text{and} \quad g'(\lambda) = \frac{I}{bd}\{\frac{\partial}{\partial\lambda}\check{g}(\lambda,t)\}.$$

Then

$$f(\lambda)g(\lambda) = \frac{I}{(bd)^2}\left\{\int_0^t \check{f}(\lambda,t-u)\check{g}(\lambda,u)\,du\right\}$$

and so

$$\begin{aligned}(f(\lambda)g(\lambda))' &= \frac{I}{(bd)^2}\left\{\frac{\partial}{\partial\lambda}\int_0^t \check{f}(\lambda,t-u)\check{g}(\lambda,u)\,du\right\} \\ &= \frac{I}{(bd)^2}\left\{\int_0^t \check{g}(\lambda,u)\frac{\partial}{\partial\lambda}\check{f}(\lambda,t-u)\,du \right. \\ &\qquad \left. + \int_0^t \check{f}(\lambda,t-u)\frac{\partial}{\partial\lambda}\check{g}(\lambda,u)\,du\right\} \\ &= \frac{I}{(bd)^2}\{\frac{\partial}{\partial\lambda}\check{f}(\lambda,t)\}\{\check{g}(\lambda,t)\} + \frac{I}{(bd)^2}\{\check{f}(\lambda,t)\}\{\frac{\partial}{\partial\lambda}\check{g}(\lambda,t)\} \\ &= f'(\lambda)g(\lambda) + f(\lambda)g'(\lambda).\end{aligned}$$

(iv) Let $f(\lambda) = c \in C/C$ for $\lambda \in (\lambda_1,\lambda_2)$. Then $c = \frac{b}{a}$ with $b \in C$, $a \in C$.

Hence $c = \frac{b}{a} = \frac{h}{ha}\{b(t)\}$, and so, taking the generalized derivative, we obtain

$$\frac{dc}{d\lambda} = \frac{h}{ha}\{\frac{\partial}{\partial\lambda}b(t)\} = 0, \quad \text{i.e.,} \quad f'(\lambda) = 0.$$

Conversely, let $f'(\lambda) = 0$. Then we have

$$f(\lambda) = \frac{a}{b}\{\hat{f}(\lambda,t)\} \quad \text{and} \quad \frac{\partial f(\lambda,t)}{\partial\lambda} = 0.$$

Hence $\{\hat{f}(\lambda,t)\}$ does not depend upon $\lambda$, so that

$$\hat{f}(\lambda,t) = \{\hat{f}(t)\}, \quad \text{i.e.,} \quad f(\lambda) = \frac{a}{b}\{\hat{f}(t)\} \in C/C.$$

## §24. EXPONENTIAL HYPERFUNCTION $\exp(\lambda s) = e^{\lambda s}$ $(-\infty < \lambda < \infty)$.

We recall that the shift operator

$$s^2\{h_1(\lambda,t)\} \quad (\lambda \geqq 0 \quad \text{and} \quad s^2\{h_1(0,t)\} = s^2\{t\} = I)$$

was denoted by

$$\exp(-\lambda s) = e^{-\lambda s}$$

$$\begin{cases} e^{-\lambda s} = \exp(-\lambda s) = \dfrac{\{h_1(\lambda,t)\}}{h^2} \in C_H \subseteq C/C \quad (\lambda \geqq 0), \\ e^{os} = I. \end{cases} \tag{24.1}$$

We shall define $e^{\lambda s}$ $(\lambda > 0)$ as follows.

PROPOSITION 20. Since $\{h_1(\lambda,t)\} \neq 0$ for any $\lambda > 0$, there exists, by (18.6), the (multiplicative) inverse operator of $e^{-\lambda s}$ given by

$$(e^{-\lambda s})^{-1} = \frac{h^2}{\{h_1(\lambda,t)\}} = \frac{I}{e^{-\lambda s}} \in C/C \quad (\lambda \geqq 0). \tag{24.2}$$

Hence, defining

$$e^{\lambda s} = (e^{-\lambda s})^{-1} \quad (\lambda \geqq 0), \tag{24.2)'}$$

we obtain

$$e^{\lambda s}e^{-\lambda s} = I \quad (\lambda \geqq 0) \tag{24.3}$$

and furthermore

$$e^{\lambda s}e^{\mu s} = e^{(\lambda+\mu)s} \quad (-\infty < \lambda,\mu < \infty). \tag{24.4}$$

PROOF OF (24.4): When $\lambda \leqq 0$ and $\mu \leqq 0$, (24.4) holds by (22.6) and $e^{os} = I$. Hence, by (24.3), we have (24.4) when $\lambda \geqq 0$ and $\mu \geqq 0$. Therefore, when $\lambda \geqq \mu \geqq 0$,

$$e^{(\lambda-\mu)s}e^{\mu s} = e^{\lambda s}, \quad \text{i.e.,} \quad e^{(\lambda-\mu)s} = e^{\lambda s}e^{-\mu s}.$$

Thus, by (24.3),

$$e^{(\lambda-\mu)s} = e^{\lambda s}e^{-\mu s} \quad \text{when} \quad 0 \leqq \lambda \leqq \mu \quad \text{as well.}$$

THEOREM 21. $f(\lambda) = \exp(\lambda s) = e^{\lambda s}$ $(-\infty < \lambda < \infty)$ does not equal zero for any value of $\lambda$, and

$$\begin{cases} f'(\lambda) = sf(\lambda) \quad (-\infty < \lambda < \infty), \quad f(0) = I, \\ \text{generalized} \lim\limits_{\lambda\to\lambda_0} f(\lambda) = f(\lambda_0) \quad \text{for} \quad -\infty < \lambda_0 < \infty. \end{cases} \tag{24.5}$$

Conversely, any hyperfunction-valued function $f(\lambda)$ satisfying (24.5) is equal to $e^{\lambda s}$.

PROOF: That $e^{\lambda s} \neq 0$ is clear from (24.3). Next, for any $\lambda$, we take $\alpha$ with $\alpha > \lambda$. Then, by (24.4),

$$e^{\lambda s} = e^{\alpha s}e^{-(\alpha-\lambda)s}. \tag{24.6}$$

Since $\alpha-\lambda > 0$,

$$\frac{de^{-(\alpha-\lambda)s}}{d(\alpha-\lambda)} = -se^{-(\alpha-\lambda)s} \quad \text{(by (23.3))},$$

i.e.,

$$\frac{de^{-(\alpha-\lambda)s}}{d\lambda} = \frac{de^{-(\alpha-\lambda)s}}{d(\alpha-\lambda)}\,\frac{d(\alpha-\lambda)}{d\lambda} = se^{-(\alpha-\lambda)s}.$$

Thus, since $e^{\alpha s} \neq 0$, we can apply (i) of Proposition 19, obtaining

$$\frac{de^{\lambda s}}{d\lambda} = e^{\alpha s}\,\frac{de^{-(\alpha-\lambda)s}}{d\lambda}$$
$$= e^{\alpha s}se^{-(\alpha-\lambda)s} = se^{\lambda s}.$$

Similarly, for any $\lambda_0$, we take $\alpha$ with $\alpha > \lambda_0$ and, by (24.6),

$$\text{generalized} \lim_{\lambda\to\lambda_0} e^{\lambda s} = e^{\alpha s} \left(\text{generalized} \lim_{\lambda\to\lambda_0} e^{-(\alpha-\lambda)s}\right)$$
$$= e^{\alpha s}e^{-(\alpha-\lambda_0)s} = e^{\lambda_0 s} \quad \text{(by (23.4)}.$$

In addition, we know already that

$$e^{os} = I \quad ((23.3)).$$

Therefore, (24.5) is completely proved.

THE PROOF OF THE CONVERSE PART OF THE THEOREM. Assume that $f(\lambda)$ satisfies (24.5). We put

$$f(\lambda) - e^{\lambda s} = x(\lambda)$$

and shall prove that $x(\lambda) = 0$ for every $\lambda$. By (ii) of Proposition 19, we have

$$x'(\lambda) = f'(\lambda) - (e^{\lambda s})' = sf(\lambda) - se^{\lambda s} = sx(\lambda). \tag{24.7}$$

Moreover,

$$x(0) = f(0) - e^{o\cdot s} = I - I = 0. \tag{24.8}$$

Now we shall prove that $x(\lambda) = 0$ from (24.7)-(24.8). To this purpose, take any real number $\mu$ and put, as in Miksiński [5], p. 191,

$$y(\lambda) = x(\lambda)x(2\mu-\lambda). \tag{24.9}$$

By (iii) of Proposition 19, we obtain

$$y'(\lambda) = x'(\lambda)x(2\mu-\lambda) - x(\lambda)x'(2\mu-\lambda)$$
$$= sx(\lambda)x(2\mu-\lambda) - sx(\lambda)x(2\mu-\lambda) = 0.$$

Hence, by (iv) of Proposition 19, there exists an element $\frac{b_1}{a_1}$ of $C/C$ such that

$$y(\lambda) = x(\lambda)x(2\mu-\lambda) = \frac{b_1}{a_1} \quad \text{for all } \lambda.$$

Hence, taking $\lambda = 0$, we obtain

$$y(0) = x(0)x(2\mu) = 0 \quad \text{since} \quad x(0) = 0.$$

This implies that $\frac{b_1}{a_1} = 0$, whence

$$y(\lambda) = x(\lambda)x(2\mu-\lambda) = 0 \quad \text{for all } \lambda. \tag{24.10}$$

Therefore, by putting $\mu = \lambda$, we have

$$x(\lambda)x(\lambda) = 0 \quad \text{for all } \lambda.$$

$x(\lambda)$ being in $C/C$, we have $x(\lambda) = \frac{b}{a}$ and so

$$\frac{b^2}{a^2} = 0, \tag{24.11}$$

$b \in C$ and $a \in C$ being dependent on the parameter $\lambda$. We must have $a \neq 0$ since it is the denominator of $\frac{b}{a}$. Hence, by Theorem 16, $a^2 \neq 0$. Thus by (24.11), $b^2 = 0$. Again by Theorem 16, $b = 0$.

This proves that $x(\lambda) = \frac{b}{a} = 0$.

REMARK 24.1. By virtue of the uniqueness, proved above, of the solution $f(\lambda)$ of (24.5), we are justified in denoting $f(\lambda)$ by $\exp(\lambda s) = e^{\lambda s}$.

COROLLARY OF THEOREM 21. By taking the generalized derivative of $f'(\lambda) = sf(\lambda)$, we obtain

$$f''(\lambda) = (f'(\lambda))' = sf'(\lambda) = s^2 f(\lambda),$$

and, more generally,

$$f^{(n)}(\lambda) = s^n f(\lambda) \quad \text{for} \quad f(\lambda) = e^{\lambda s}. \tag{24.12}$$

## §25. EXAMPLES OF GENERALIZED LIMIT. POWER SERIES IN $e^{\lambda s}$

The reader is referred to Remark 23.2.

PROPOSITION 21. Let

$$\{f_n \in C/C;\ n = 1,2,\ldots\}$$

be a sequence of hyperfunctions. For a certain $\frac{b}{a} \neq 0$ of $C/C$, let there exist a sequence of functions $\{\hat{f}_n(t)\} \in C$ $(n = 1,2,\ldots)$ and a function $\{\hat{f}(t)\} \in C$ in such a way that

$$\frac{b}{a} f_n = \{\hat{f}_n(t)\} \text{ and } \lim_{n\to\infty} \hat{f}_n(t) = \hat{f}(t) \text{ uniformly on every finite closed interval } [0,T], \ 0 < T < \infty. \tag{25.1}$$

Then we shall call

$$\frac{a}{b}\{\lim_{n\to\infty} \hat{f}_n(t)\} = \frac{a}{b}\{\hat{f}(t)\} \tag{25.2}$$

the *generalized limit* of the sequence $\{f_n \in C/C;\ n = 1,2,\ldots\}$.

PROOF OF THE CONSISTENCY OF THE ABOVE DEFINITION: For another $\frac{b_1}{a_1} \neq 0$ of $C/C$, let there exist a sequence of functions $\frac{b_1}{a_1} f_n = \{\check{f}_n(t)\} \in C$ $(n = 1,2,\ldots)$ and a function $\{\check{f}(t)\} \in C$ in such a way that $\lim_{n\to\infty} \check{f}_n(t) = \check{f}(t)$ uniformly on every finite closed interval $[0,T]$. Then we can prove

$$\frac{a}{b}\{\hat{f}(t)\} = \frac{a_1}{b_1}\{\check{f}(t)\}.$$

In fact, we have

$$f_n = \frac{a}{b}\{\hat{f}_n(t)\} = \frac{a_1}{b_1}\{\check{f}_n(t)\},$$

so that

$$ab_1\hat{f}_n = a_1b\,\check{f}_n,$$

i.e.,

$$\int_0^t (ab_1)(t-u)\hat{f}_n(u)du = \int_0^t (a_1b)(t-u)\check{f}_n(u)du.$$

Thus, by letting $n \to \infty$, we obtain

$$\int_0^t (ab_1)(t-u)\hat{f}(u)du = \int_0^t (a_1b)(t-u)\check{f}(u)du,$$

i.e.,

$$ab_1\hat{f} = a_1b\,\check{f}.$$

Therefore, by multiplying by $\frac{I}{bb_1}$, we have

$$\frac{a}{b}\hat{f} = \frac{a_1}{b_1}\check{f}.$$

EXAMPLES OF GENERALIZED LIMIT.

EXAMPLE 25.1. For $f = \{f(t)\} \in C$, we have

$$f^1 = f, \quad f^2 = f*f, \ldots, f^n = f*f^{n-1} = f^{n-1}*f \quad (n = 1,2,\ldots;\ f^0 = I).$$

If $f_n$ denotes $f^n$, we obtain

$$\text{generalized } \lim_{n\to\infty} f_n = 0. \tag{25.3}$$

PROOF: Choose $T > 0$ and let $0 \leqq t \leqq T$. Then $\max_{0\leqq t\leqq T} |f(t)| \leqq M$ for some finite $M > 0$. Assuming

$$|f^n(t)| \leqq \frac{M^n t^{n-1}}{(n-1)!} \quad (0 \leqq t \leqq T;\ n = 1,2,\ldots,k) \tag{25.4}$$

we obtain

$$|f^{k+1}(t)| = \left|\int_0^t f(t-u) f^k(u)\,du\right|$$

$$\leqq \int_0^t M \frac{M^k u^{k-1}}{(k-1)!}\,du = \frac{M^{k+1} t^k}{k!} \quad (0 \leqq t \leqq T)$$

Hence (25.4) holds for all positive integer $k$.

Therefore, if we take a positive integer $n_0$ satisfying

$MT/n_0 \leqq 1/2$,

then we obtain for $k > n_0$,

$$|f^{k+1}(t)| \leqq \frac{M^{n_0} T^{n_0}}{n_0!} \left(\frac{MT}{n_0}\right)^{k+1-n_0} \times \frac{1}{T}$$

$$\leqq \frac{M^{n_0} T^{n_0-1}}{n_0!} \times \left(\frac{1}{2}\right)^{k+1-n_0} \quad (0 \leqq t \leqq T)$$

This proves (25.3). In this example, we have made use of $\frac{b}{a} = I$.

EXAMPLE 25.2. For $f_n = \{\sin nt\}$, we have

$$f_n = s\left\{\int_0^t \sin nu\,du\right\} = s\left\{\frac{1-\cos nt}{n}\right\}.$$

Hence

$$\text{generalized } \lim_{n\to\infty} \{\sin nt\} = s\left\{\lim_{n\to\infty} \frac{1-\cos nt}{n}\right\} = s\{0\} = 0. \tag{25.5}$$

EXAMPLE 25.3. For $f_n = (\frac{s}{n} - I) = s^2\left\{\frac{h}{n} - h^2\right\}$, we have

$$\text{generalized } \lim_{n\to\infty} (\frac{s}{n} - I) = s^2\{\lim_{n\to\infty} (\frac{h}{n} - h^2)\} = s^2(-h^2) = -I. \tag{25.6}$$

EXAMPLE 25.4. For $f_n = (\frac{s}{n} - I)^{-1} = \frac{I}{s/n - I} = \frac{nI}{s - nI} = \{ne^{nt}\}$,

$$\text{generalized } \lim_{n\to\infty} \{ne^{nt}\} \text{ does not exist.} \tag{25.7}$$

PROOF: Assume that $\frac{b}{a} \neq 0$ of $C/C$ exists such that $\frac{b}{a}\{ne^{nt}\} = \{\hat{f}_n(t)\} \in C$ satisfies

$$\lim_{n\to\infty} \hat{f}_n(t) = \hat{f}(t) \in C \text{ exists uniformly on } [0,T] \text{ for every positive } T < \infty. \tag{25.8}$$

From this we shall derive a contradiction, as follows. (25.8) implies that the sequence of functions

$$a\{\hat{f}_n(t)\} = b\{ne^{nt}\} = n\left\{\int_0^t b(t-u)e^{nu}du\right\}$$

converges uniformly on every finite interval $[0,T]$ as $n \to \infty$. Hence, by taking $t = T > 0$, we see that

$$\text{a finite } \lim_{n\to\infty} n\int_0^T b(T-u)e^{nu}du \text{ must exist.}$$

Thus there must exist a constant $M_T$ such that

$$\sup_{n=1,2,\ldots} \left|\int_0^T b(T-u)e^{nu}du\right| \leqq M_T < \infty. \tag{25.9}$$

Therefore, by Mikusiński's theorem of moments (Theorem 20), we must have

$$b(T-u) = 0 \text{ for all } u \text{ in } [0,T], \text{ i.e., } b(t) \equiv 0 \text{ in } [0,T].$$

Since $T > 0$ was arbitrarily chosen, we must have $b(t) \equiv 0$ in $[0,\infty)$. This contradicts the assumption $\frac{b}{a} \neq 0$.

EXAMPLE 25.5. Let $\{f(t)\} \in C$ satisfy $f(t) \geqq 0$ and $\int_0^\infty f(u)du = 1$. Then

$$\text{generalized } \lim_{n\to\infty} \{nf(nt)\} = I. \tag{25.10}$$

PROOF:

$$\{nf(nt)\} = s^2h^2\{nf(nt)\} = s^2h\left\{\int_0^{nt} f(v)dv\right\} = s^2\left\{\int_0^t du \int_0^{nu} f(v)dv\right\}.$$

On the other hand, we have, for every $T > 0$,

$$\lim_{n\to\infty}\int_0^t du \int_0^{nu} f(v)dv = t \quad \text{uniformly on} \quad [0,T].$$

In fact, by $\int_0^\infty f(v)dv = 1$,

$$\lim_{n\to\infty}\left[t - \int_0^t dt \int_0^{nt} f(v)dv\right] = \lim_{n\to\infty}\int_0^t\left[1 - \int_0^{nu} f(v)dv\, du\right]$$
$$= \lim_{n\to\infty}\int_0^t\left[\int_{nu}^\infty f(v)dv\right]du = 0.$$

This proves (25.10).

EXAMPLE 25.6. For any complex number $\beta$ and any positive number $\lambda$, we have

$$(I - \beta e^{-\lambda s})^{-1} = s^2\{f_\beta(\lambda,t)\} \tag{25.11}$$

where the complex-valued function $f_\beta(\lambda) = \{f_\beta(\lambda,t)\}$ is given as follows:

For $T \geqq \lambda$, let $k$ be the positive integer satisfying $k\lambda \leqq T < (k+1)\lambda$. Then $\{f_\beta(\lambda,t)\} = h^2 + \beta\{h_1(\lambda,t)\}$ (25.12)
$+ \beta^2\{h_1(2\lambda,t)\} + \ldots + \beta^k\{h_1(k\lambda,t)\}$, where $0 \leqq t \leqq T$.

PROOF: Let $T \geqq \lambda$. Then, for $0 \leqq t \leqq T$,

$$s^2\{f_\beta(\lambda,t)\} = I + \beta e^{-\lambda s} + \beta^2 e^{-2\lambda s} + \cdots + \beta^k e^{-k\lambda s}$$

$$(k\lambda \leqq T < (k+1)\lambda;\ k = 1,2,\ldots).$$

On the other hand,

$$(I-\beta e^{-\lambda s})(I + \beta e^{-\lambda s} + \beta^2 e^{-2\lambda s} + \cdots + \beta^k e^{-k\lambda s})$$
$$= (I + \beta e^{-\lambda s} + \beta^2 e^{-2\lambda s} + \cdots + \beta^k e^{-k\lambda s})(I-\beta e^{-\lambda s})$$
$$= I - \beta^{k+1} e^{-(k+1)\lambda s}$$

and

$$\text{generalized} \lim_{n\to\infty} \beta^n e^{-n\lambda s} = s^2\left\{\lim_{n\to\infty} \beta^n h^2 e^{-n\lambda s}\right\}$$
$$= s^2\left\{\lim_{n\to\infty} \beta^n h_1(n\lambda,t)\right\} = s^2 0 = 0,$$

because, for $0 \leqq t < T < (k+1)\lambda$,

$$h_1(n\lambda,t) = 0 \quad \text{if} \quad n \geqq (k+1).$$

Therefore, we obtain

$$
\begin{aligned}
s^2\{f_\beta(\lambda,t)\} &= \text{generalized } \lim_{k\to\infty} (I + \beta e^{-\lambda s} + \cdots + \beta^k e^{-k\lambda s}) \\
&= I + \beta e^{-\lambda s} + \cdots + \beta^k e^{-k\lambda s} + \cdots \\
&= (I - \beta e^{-\lambda s})^{-1} \quad (\lambda > 0).
\end{aligned} \tag{25.13}
$$

REMARK 25.1. The existence of the multiplicative inverse $(I - \beta e^{-\lambda s})^{-1}$ of $(I - \beta e^{-\lambda s}) \in C/C$ is proved directly from $(I - \beta e^{-\lambda s}) \neq 0$. This inequality is clear when $\beta = 0$; and if $\beta \neq 0$, then the equality $I = \beta e^{-\lambda s}$ implies the contradiction $\beta^{-1}h^2 = h^2 e^{-\lambda s} = \{h_1(\lambda,t)\}$ $(\lambda > 0)$.

EXAMPLE 25.7. Let $\lambda > 0$ and let $f = \{f(t)\} \in K[0,\infty)$ vanish for $\lambda < t < \infty$. Then

$$
\begin{cases}
g = \{g(t)\} = (I - e^{-\lambda s})^{-1} f \\
\quad = f + e^{-\lambda s} f + e^{-2\lambda s} f + \cdots + e^{-k\lambda s} f + \cdots
\end{cases} \tag{25.14}
$$

is periodic with period $\lambda$, i.e.,

$$g(t + \lambda) = g(t).$$

PROOF: As was discussed in §22, we have

$$
e^{-k\lambda s} f = \begin{cases} 0, & 0 \leqq t < k\lambda \\ f(t-k\lambda), & 0 < k\lambda < t \end{cases} \in K.
$$

REMARK 25.2. If $g \in K[0,\infty)$ is periodic with period $\lambda > 0$, then $g$ can be represented as

$$
\begin{cases}
g = (I - e^{-\lambda s})^{-1} f, \quad \text{where } f = (I - e^{-\lambda s}) g \in K[0,\infty) \\
\text{and } f(t) = 0 \quad \text{for } \lambda < t < \infty.
\end{cases} \tag{25.15}
$$

A CONCRETE EXAMPLE OF (22.15). Let $\lambda > 0$ and put

$$
\begin{aligned}
f = \{f(t)\} &= h^2 - e^{-\lambda s}(h^2 + \lambda h) \\
&= \{t\} - \begin{Bmatrix} 0, & 0 \leqq t < \lambda \\ t+\lambda-\lambda, & 0 < \lambda \leqq t \end{Bmatrix} = \begin{Bmatrix} t, & 0 \leqq t < \lambda \\ 0, & 0 < \lambda \leqq t \end{Bmatrix}.
\end{aligned}
$$

Thus

$$
\begin{cases}
f = \dfrac{I}{s^2} - \left(\dfrac{I}{s^2} + \dfrac{[\lambda]}{s}\right) e^{-\lambda s}, \\
g = (I - e^{-\lambda s})^{-1} f
\end{cases} \tag{25.16}
$$

gives an example of (22.15). (See the saw-toothed graph below.)

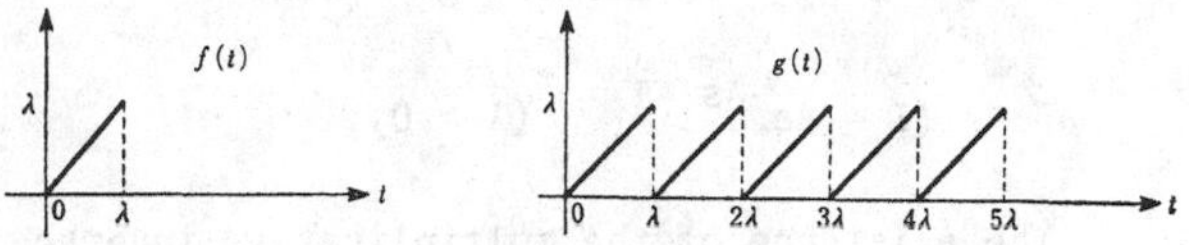


§26. $\int_0^\infty e^{-\lambda s} f(\lambda)d\lambda = \{f(t)\}$ FOR $\{f(t)\} \in C$. Let $F(\lambda)$ be a hyperfunction-valued function $(\lambda_1 < \lambda < \lambda_2)$. Then the *generalized integral* $\int_{\lambda_1}^{\lambda_2} F(\lambda)d\lambda$ is defined as follows. For a certain $\frac{b}{a} \neq 0$ of $C/C$, $\frac{b}{a} F(\lambda) = \{\hat{F}(\lambda,t)\}$ is a complex-valued continuous function defined on $\mathcal{D} = \{\{\lambda,t\}: \lambda_1 < \lambda < \lambda_2$ and $0 \leqq t\}$ such that $\int_{\lambda_1}^{\lambda_2} \hat{F}(\lambda,t)d\lambda \in C$ as a function of $t$. Then, by definition, we have

$$\int_{\lambda_1}^{\lambda_2} F(\lambda)d\lambda = \frac{a}{b}\left\{\int_{\lambda_1}^{\lambda_2} \hat{F}(\lambda,t)d\lambda\right\} \tag{26.1}$$

REMARK 26.1. The above definition of the generalized integral is *reasonable* in the following sense: If another $\frac{b_1}{a_1} \neq 0$ of $C/C$ exists in such a way that $\{\hat{\hat{F}}(\lambda,t)\} = \frac{b_1}{a_1} F(\lambda)$ is continuous on $\mathcal{D}$ and $\left\{\int_{\lambda_1}^{\lambda_2} \hat{\hat{F}}(\lambda,t)d\lambda\right\} \in C$, then we have

$$\frac{a}{b}\left\{\int_{\lambda_1}^{\lambda_2} \hat{F}(\lambda,t)d\lambda\right\} = \frac{a_1}{b_1}\left\{\int_{\lambda_1}^{\lambda_2} \hat{\hat{F}}(\lambda,t)d\lambda\right\}. \tag{26.2}$$

PROOF OF (26.2): From

$$F(\lambda) = \frac{a}{b}\{\hat{F}(\lambda,t)\} = \frac{a_1}{b_1}\{\hat{\hat{F}}(\lambda,t)\}$$

we obtain

$$ab_1\{\hat{F}(\lambda,t)\} = a_1 b\{\hat{\hat{F}}(\lambda,t)\},$$

and so, by integrating,

$$\int_{\lambda_1}^{\lambda_2} d\lambda\left(\int_0^t ab_1(t-u)\hat{F}(\lambda,u)du\right) = \int_{\lambda_1}^{\lambda_2} d\lambda\left(\int_0^t a_1b(t-u)\hat{\hat{F}}(\lambda,u)du\right).$$

Thus

$$ab_1\left\{\int_{\lambda_1}^{\lambda_2} \hat{F}(\lambda,t)d\lambda\right\} = a_1b\left\{\int_{\lambda_1}^{\lambda_2} \hat{\hat{F}}(\lambda,t)d\lambda\right\}$$

and, by multiplying by $\frac{I}{b_1b}$, we obtain (26.2).

THEOREM 22. Let $\{f(t)\} \in C$. Then

$$\int_0^\infty e^{-\lambda s} f(\lambda)d\lambda = \{f(t)\}. \tag{26.3}$$

PROOF: For $0 \leqq \lambda < \infty$ and $0 \leqq t < \infty$, we have

$$F(\lambda) = e^{-\lambda s} f(\lambda) = s^2h^2e^{-\lambda s} f(\lambda)$$

$$= s^2\{h_1(\lambda,t)f(\lambda)\} = s^2\begin{Bmatrix} 0, & 0 \leqq t < \lambda \\ (t-\lambda)f(\lambda), & 0 \leqq \lambda \leqq t \end{Bmatrix}.$$

Hence, by Theorem 2,

$$\int_0^\infty e^{-\lambda s} f(\lambda)d\lambda = s^2\left\{\int_0^\infty h_1(\lambda,t)f(\lambda)d\lambda\right\}$$

$$= s^2\left\{\int_0^t (t-\lambda)f(\lambda)d\lambda\right\}$$

$$= s\left\{(t-t)f(t) + \int_0^t \frac{\partial(t-\lambda)}{\partial t} f(\lambda)d\lambda + \int_0^0 (0-\lambda)f(\lambda)d\lambda\right\}$$

$$= s\left\{\int_0^t f(\lambda)d\lambda\right\} = \left\{f(t) + \int_0^0 f(\lambda)d\lambda\right\} = \{f(t)\}.$$

REMARK 26.2. The above integral $\int_0^\infty e^{-\lambda s} f(\lambda)d\lambda$ might be called the *generalized Laplace integral of* $f(\lambda)$. In Heaviside's operational calculus,* the operator of differentiation was denoted by $p$. The interpretation of the operators like $p$ as given by Heaviside is difficult to justify, and the range of the validity of his calculus remains unclear even now, although it was widely noticed that his calculus gives correct

*Oliver Heaviside: Electromagnetic Theory, London, 1-3 (1893-1899).

results in general.* Thus G. Doetsch** and many other mathematicians strove for the foundation of Heaviside's calculus by virtue of the theory of the Laplace transform

$$\int_0^\infty e^{-pt}f(t)dt \quad \text{or} \quad p\int_0^\infty e^{-pt}f(t)dt.$$

However, the use of such integrals naturally confronts restriction due to growth behavior of the numerical function $f(t)$ as $t \to \infty$. See Remark 10.2.

## §27. EXPONENTIAL HYPERFUNCTION $\exp(-\lambda s^{1/2}) = e^{-\lambda s^{1/2}}$ $(-\infty < \lambda < \infty)$.

PROPOSITION 22. The hyperfunction-valued function $f(\lambda)$ given by

$$\begin{cases} f(\lambda,t) = \dfrac{\lambda}{2\sqrt{\pi t^3}}\, e^{-\lambda^2/4t} \quad (\lambda > 0,\ t \geqq 0), \\ f(\lambda,0) = 0 \quad (\lambda > 0), \end{cases} \tag{27.1}$$

satisfies

$$\begin{cases} \dfrac{d}{d\lambda}\{f(\lambda,t)\} = -s^{1/2}\{f(\lambda,t)\} \in C_H \subsetneqq C/C, \\ \text{generalized } \lim\limits_{\lambda\downarrow 0} \{f(\lambda,t)\} = I. \end{cases} \tag{27.2}$$

PROOF: We have, by (13.3),

$$h^{1/2}\{f(\lambda,t)\} = \left\{\int_0^t \frac{\lambda}{2\pi}(t-u)^{-1/2}u^{-3/2}\exp(-\lambda^2/4u)\,du\right\}.$$

By the substitution

$$u = \left(\frac{1}{t} + \frac{4\sigma^2}{\lambda^2}\right)^{-1}$$

we change integration variables, from $u$ to $\sigma$. Then

$$(t-u)^{-1/2}u^{-3/2}$$

$$= \left(t - 1\Big/\left(\frac{1}{t} + \frac{4\sigma^2}{\lambda^2}\right)\right)^{-1/2}\left(1\Big/\left(\frac{1}{t} + \frac{4\sigma^2}{\lambda^2}\right)\right)^{-3/2}$$

$$= \left(\left(1 + \frac{4\sigma^2 t}{\lambda^2} - 1\right)\Big/\left(\frac{1}{t} + \frac{4\sigma^2}{\lambda^2}\right)\right)^{-1/2}\left(1\Big/\left(\frac{1}{t} + \frac{4\sigma^2}{\lambda^2}\right)\right)^{-3/2}$$

---

* In 1921, Heaviside received the first Faraday Medal of the Institution of Electrical Engineers of Great Britain.

** Gustav Doetsch (1892-1977); Einführung in Theorie und Anwendung der Laplace-Transformation, Birkhäuser Verlag (1958).

$$= \frac{\lambda}{2\sigma t^{1/2}}\left(\frac{1}{t} + \frac{4\sigma^2}{\lambda^2}\right)^2,$$

$$\exp\left(-\frac{\lambda^2}{4u}\right) = \exp\left(-\frac{\lambda^2}{4t} - \sigma^2\right) = \exp\left(-\frac{\lambda^2}{4t}\right)e^{-\sigma^2},$$

$$du = -\left(\frac{8\sigma}{\lambda^2}\left(\frac{1}{t} + \frac{4\sigma^2}{\lambda^2}\right)^2\right)d\sigma.$$

Hence

$$h^{1/2}\{f(\lambda,t)\} = \int_\infty^0 -\frac{\lambda}{2\pi}\frac{\lambda}{2\sigma t^{1/2}}\frac{8\sigma}{\lambda^2}\exp\left(-\frac{\lambda^2}{4t}\right)e^{-\sigma^2}d\sigma$$

and so, by $\int_0^\infty e^{-\sigma^2}d\sigma = \sqrt{\pi}/2$ ((12.8)), we obtain

$$h^{1/2}\{f(\lambda,t)\} = \frac{I}{\sqrt{s}}\{f(\lambda,t)\} = \left\{\frac{1}{\sqrt{\pi t}}e^{-\lambda^2/4t}\right\}. \tag{27.3}$$

Thus

$$\begin{aligned}\{f(\lambda,t)\} &= \frac{h}{h^{3/2}}\left\{\frac{1}{\sqrt{\pi t}}e^{-\lambda^2/4t}\right\} \\ &= s^{3/2}\left\{\int_0^t \frac{1}{\sqrt{\pi u}}e^{-\lambda^2/4u}du\right\}.\end{aligned} \tag{27.4}$$

Hence the generalized derivative $\frac{d}{d\lambda}\{f(\lambda,t)\}$ is given by

$$\begin{aligned}\frac{d}{d\lambda}\{f(\lambda,t)\} &= s^{3/2}\left\{\int_0^t \frac{\partial}{\partial\lambda}\frac{1}{\sqrt{\pi u}}\exp(-\lambda^2/4u)du\right\} \\ &= s^{1/2}s\left\{\int_0^t \frac{-1}{\sqrt{\pi u}}\frac{\lambda}{2u}\exp(-\lambda^2/4u)du\right\} \\ &= -s^{1/2}\left\{f(\lambda,t) + \int_0^0 \frac{\lambda}{2\sqrt{\pi u^3}}\exp(-\lambda^2/4u)du\right\} \\ &= -s^{1/2}\{f(\lambda,t)\}.\end{aligned}$$

Furthermore, by (27.4) and (13.3),

$$\begin{aligned}&\text{generalized} \lim_{\lambda\downarrow 0} f(\lambda) \\ &= \lim_{\lambda\downarrow 0}\{f(\lambda,t)\} = s^{3/2}\left\{\lim_{\lambda\downarrow 0}\int_0^t \frac{1}{\sqrt{\pi u}}\exp(-\lambda^2/4u)du\right\} \\ &\qquad = s^{3/2}\left\{\int_0^t \frac{1}{\sqrt{\pi u}}du\right\} = s^{3/2}\cdot h^{1/2}\cdot h = s^{3/2}\cdot h^{3/2} = I.\end{aligned}$$

Therefore, by putting $f(0) = I$, we have

$$\begin{cases} f'(\lambda) = -s^{1/2}f(\lambda) \quad (\lambda > 0), \quad f(0) = I, \text{ and so} \\ \text{generalized } \lim\limits_{\lambda\to\lambda_0} f(\lambda) = f(\lambda_0) \quad (\lambda_0 \geqq 0). \end{cases} \tag{27.2)'$$

DEFINITION 27.1.* We shall denote

$$\begin{cases} \exp(-\lambda s^{1/2}) = e^{-\lambda s^{1/2}} = \left\{\dfrac{\lambda}{2\sqrt{\pi t^3}} \exp(-\lambda^2/4t)\right\} \quad (\lambda > 0), \\ \qquad = I \quad (\lambda = 0). \end{cases} \tag{27.5}$$

Then, by (27.2)',

$$\begin{cases} \dfrac{d}{d\lambda} \exp(-\lambda s^{1/2}) = -s^{1/2}\exp(-\lambda s^{1/2}) \quad (\lambda > 0), \\ \text{generalized } \lim\limits_{\lambda\downarrow 0} \exp(-\lambda s^{1/2}) = I = \exp(-0\cdot s^{1/2}). \end{cases} \tag{27.6}$$

PROPOSITION 23. If $\mu > \lambda > 0$, then

$$\exp(-\lambda s^{1/2})\exp(-(\mu-\lambda)s^{1/2}) = \exp(-\mu s^{1/2}). \tag{27.7}$$

PROOF: Putting $f(\lambda) = \exp(-\lambda s^{1/2})$, we shall prove

$$f(\lambda)f(\mu-\lambda) = f(\mu).$$

By (iii) of Proposition 19 (§23), we have

$$\begin{aligned} \frac{d}{d\lambda}(f(\lambda)f(\mu-\lambda)) &= f'(\lambda)f(\mu-\lambda) + f(\lambda)\frac{d}{d\lambda} f(\mu-\lambda) \\ &= -s^{1/2}f(\lambda)f(\mu-\lambda) + f(\lambda)s^{1/2}f(\mu-\lambda) = 0 \end{aligned}$$

because

$$\frac{d}{d\lambda} f(\mu-\lambda) = \frac{d}{d(\mu-\lambda)} f(\mu-\lambda)\cdot \frac{d(\mu-\lambda)}{d\lambda} .$$

Hence, by (iv) of Proposition 19, we have

$$f(\lambda)f(\mu-\lambda) = \frac{b}{a} \in C/C$$

so that $\frac{b}{a}$ does not depend on $\lambda$. Thus

$$\begin{aligned} f(\lambda)f(\mu-\lambda) &= \text{generalized } \lim_{\lambda\downarrow 0} f(\lambda)f(\mu-\lambda) \\ &= If(\mu) = f(\mu). \end{aligned}$$

*Due to J. Mikusiński [5], p. 220.

DEFINITION 27.2. We shall denote

$$\exp(\lambda s^{1/2}) = \exp(-\lambda s^{1/2})^{-1} = \frac{I}{\exp(-\lambda s^{1/2})} \quad (\lambda \geqq 0). \tag{27.8}$$

The existence of the (multiplicative) inverse $\exp(-\lambda s^{1/2})^{-1}$ of $\exp(-\lambda s^{1/2})$ for $\lambda > 0$ is proved as follows:

Since $s^{3/2} \in C_H$ ((13.9)), we have, by (27.1), (27.4) and (27.5),

$$\exp(-\lambda s^{1/2}) = s^{3/2}\left\{\int_0^t \frac{1}{\sqrt{\pi u}} \exp(-\lambda^2/4u)\,du\right\} \in C/C \quad (\lambda > 0).$$

Moreover, this expression shows that $\exp(-\lambda s^{1/2}) \neq 0$ for every $\lambda > 0$. Indeed, if it were otherwise, then by (27.4),

$$h^{3/2}\exp(-\lambda s^{1/2}) = \left\{\int_0^t \frac{1}{(\pi u)^{1/2}} \exp(-\lambda^2/4u)\,du\right\} = 0$$

for all $t \geq 0$. We shall use Proposition 12.

Thus, if $\lambda > 0$, there exists the (multiplicative) inverse

$\exp(-\lambda s^{1/2})^{-1}$ of $\exp(-\lambda s^{1/2})$ in the ring $C/C$.

We shall denote, by using $\exp(-0s^{1/2}) = I$,

$$\exp(-\lambda s^{1/2})^{-1} = \frac{I}{\exp(-\lambda s^{1/2})} = \exp(\lambda s^{1/2}) \quad (\lambda \geqq 0).$$

We have thus proved (27.8) and

$$\exp(\lambda s^{1/2})\exp(-\lambda s^{1/2}) = I \quad (\lambda \geqq 0). \tag{27.8)'$$

THEOREM 23. We have

$$\exp(\lambda s^{1/2})\exp(\mu s^{1/2}) = \exp((\lambda+\mu)s^{1/2}) \quad (-\infty < \lambda,\ \mu < +\infty) \tag{27.9}$$

and

$$\begin{cases} \dfrac{d}{d\lambda}\exp(\lambda s^{1/2}) = s^{1/2}\exp(\lambda s^{1/2}) & (-\infty < \lambda < +\infty), \\ \text{generalized} \lim\limits_{\lambda \to \lambda_0} \exp(\lambda s^{1/2}) = \exp(\lambda_0 s^{1/2}) & (-\infty < \lambda_0 < \infty). \end{cases} \tag{27.10}$$

We also have

$$\frac{d^n}{d\lambda^n}\exp(\lambda s^{1/2}) = (s^{1/2})^n \exp(\lambda s^{1/2}) \quad (-\infty < \lambda < \infty;\ n = 1,2,\dots) \tag{27.10)'$$

and

$$\exp(\lambda s^{1/2}) \neq 0 \qquad (-\infty < \lambda < \infty). \tag{27.11}$$

PROOF: (27.9) is a consequence of Proposition 23 and (27.8)'. The reader may follow the proof of (24.4).

(27.10) is also a consequence of (27.2)' and (27.9). The reader may follow the proof of (24.5).

Finally, (27.11) is a consequence of (27.8)'.

REMARK 27.1. We can prove that any solution $f(\lambda)$ of

$$\begin{cases} \dfrac{df(\lambda)}{d\lambda} = s^{1/2}f(\lambda) \qquad (-\infty < \lambda < \infty), \quad f(0) = I, \\ \text{generalized} \lim\limits_{\lambda\to\lambda_0} f(\lambda) = f(\lambda_0) \end{cases} \tag{27.12}$$

must coincide with $\exp(\lambda s^{1/2})$. The reader may follow the proof of the converse part of Theorem 21 (§24).

COROLLARY OF PROPOSITION 22. We have

$$\frac{I}{\sqrt{s}} \exp(-\lambda s^{1/2}) = \left\{\frac{1}{\sqrt{\pi t}} \exp(-\lambda^2/4t)\right\}, \tag{27.13}$$

$$\frac{I}{s} \exp(-\lambda s^{1/2}) = \left\{\text{Cerf } \frac{\lambda}{2\sqrt{t}}\right\}^{*}. \tag{27.14}$$

PROOF: (27.13) is already proved, by (27.1), (27.3), and (27.5). Next, we obtain from (27.5)

$$\frac{I}{s} \exp(-\lambda s^{1/2}) = h \exp(-\lambda s^{1/2})$$

$$= \left\{\int_0^t \frac{\lambda}{2\sqrt{\pi u^3}} \exp(-\lambda^2/4u)\,du\right\}.$$

Then perform the change of the variable $u \to \sigma = \dfrac{\lambda}{2\sqrt{u}}$. We have thus proved (27.14).

REMARK 27.2. For $f(\lambda) = \exp(-\lambda s^{1/2})$, we have

$$f'(\lambda) = -s^{1/2}f(\lambda),$$

$$f''(\lambda) = (f'(\lambda))' = -s^{1/2}f'(\lambda) = s^{1/2}s^{1/2}f(\lambda)$$

$$= sf(\lambda).$$

---

*Cerf $t = 2(\pi)^{-1/2} \int_t^\infty e^{-u^2}du = 1 - \text{Erf } t$, Erf $t = 2(\pi)^{-1/2} \int_0^t e^{-u^2}du$.

Since $s$ is the operator of differentiation with respect to $t$, $\exp(-\lambda s^{1/2})$ is intimately related to the heat equation

$$z_{\lambda\lambda} = z_t.$$

Thus $\exp(-\lambda s^{1/2})$ is sometimes called the *heat operator* or the *diffusion operator*.

## §28. LOGARITHMIC HYPERFUNCTION w AND EXPONENTIAL HYPERFUNCTION exp(λw)

DEFINITION 28.1. Let a hyperfunction-valued function $f(\lambda)$ satisfy

$$f'(\lambda) = wf(\lambda) \quad (-\infty < \lambda < \infty), \quad f(0) = I, \tag{28.1}$$

where $w \in C/C$. Then we call $w$ a *logarithmic hyperfunction* and $f(\lambda)$ an *exponential hyperfunction*.

EXAMPLES. $s$ and $s^{1/2}$ are logarithmic hyperfunctions.

PROPOSITION 24. Let $w$ be a logarithmic hyperfunction. Then the solution $f(\lambda)$ of (28.1) is uniquely determined, and we have

$$\text{generalized} \lim_{\lambda \to \lambda_0} f(\lambda) = f(\lambda_0) \quad (-\infty < \lambda_0 < \infty), \tag{28.2}$$

$$f(\lambda) \neq 0 \quad (-\infty < \lambda < \infty), \tag{28.3}$$

$$f(\lambda)f(\mu) = f(\lambda+\mu) \quad (-\infty < \lambda, \mu < \infty). \tag{28.4}$$

PROOF: The uniqueness of the solution $f(\lambda)$ of (28.1) is proved similarly as in the converse part of Theorem 21 (§24).*

PROOF OF (28.2): The assumption that $f(\lambda)$ has the generalized derivative $f'(\lambda)$ implies that $f(\lambda)$ is continuous in the generalized sense (see Remark 23.2) so that (28.2) is true.

PROOF OF (28.3): Assume that $f(\lambda_0) = 0$ for a certain $\lambda_0$. Then $\hat{f}(\lambda) = f(\lambda+\lambda_0)$ is a solution of

$$\hat{f}'(\lambda) = w\hat{f}(\lambda) \quad (-\infty < \lambda < \infty), \quad \hat{f}(0) = 0.$$

Hence the proof of the converse part of Theorem 21 can be modified to prove that $\hat{f}(\lambda) \equiv 0$ so that $f(\lambda) \equiv 0$.

PROOF OF (28.4): The proof of (27.9) can be modified to show that (28.4) is true.

COROLLARY. Because of the uniqueness of the solution $f(\lambda)$ of (28.1), we are justified in denoting it by $\exp(\lambda w)$.

---
* In the proof, we make use of (28.2).

PROPOSITION 25. $w = is = \sqrt{-1}\, s \in C/C$ is not a logarithmic hyperfunction.

PROOF:* Assume that there exists an $f(\lambda)$ which satisfies

$$f'(\lambda) = is\, f(\lambda) = \sqrt{-1}\, s\, f(\lambda) \quad (-\infty < \lambda < \infty), \quad f(0) = I. \tag{28.5}$$

Then there exist a closed interval $[\lambda_1, \lambda_2]$ (with $\lambda_1 < 0 < \lambda_2$) together with an element $\frac{b}{a} \neq 0$ of $C/C$ such that

$$\begin{cases} \text{both } \{\hat{f}(\lambda,t)\} = \hat{f}(\lambda) = \frac{b}{a} f(\lambda) \text{ and } \{\frac{\partial}{\partial\lambda}\hat{f}(\lambda,t)\} \\ \text{are continuous on } \mathcal{D} = \{\{\lambda,t\};\ \lambda_1 \leqq \lambda \leqq \lambda_2,\ 0 \leqq t\}, \end{cases}$$

and so the generalized derivative

$$f'(\lambda) = \frac{a}{b}\hat{f}'(\lambda) = \frac{a}{b}\{\frac{\partial}{\partial\lambda}\hat{f}(\lambda,t)\} = is\, f(\lambda)$$
$$= is\,\frac{a}{b}\hat{f}(\lambda) = is\,\frac{a}{b}\{\hat{f}(\lambda,t)\}.$$

Next let us put

$$\omega = \frac{2\pi}{\lambda_2 - \lambda_1}.$$

Then, by $f'(\lambda) = is\, f(\lambda)$, we have

$$\int_{\lambda_1}^{\lambda_2} e^{-in\omega\lambda}\hat{f}'(\lambda)d\lambda = is\int_{\lambda_1}^{\lambda_2} e^{-in\omega\lambda}\hat{f}(\lambda)d\lambda \quad (n = 1,2,\ldots). \tag{28.6}$$

The left hand side is, by integrating by parts and using $e^{-in\omega\lambda_2} = e^{-in\omega\lambda_1}$, equal to

$$e^{-in\omega\lambda_2}(\hat{f}(\lambda_2) - \hat{f}(\lambda_1)) + in\omega\int_{\lambda_1}^{\lambda_2} e^{-in\omega\lambda}\hat{f}(\lambda)d\lambda.$$

Hence, by (28.6), we have

$$e^{-in\omega\lambda_2}(\hat{f}(\lambda_2) - \hat{f}(\lambda_1)) = i(s - n\omega)\int_{\lambda_1}^{\lambda_2} e^{-in\omega\lambda}\hat{f}(\lambda)d\lambda,$$

and so

$$\frac{\hat{f}(\lambda_2) - \hat{f}(\lambda_1)}{s - n\omega} = ie^{in\omega\lambda_2}\int_{\lambda_1}^{\lambda_2} e^{-in\omega\lambda}\hat{f}(\lambda)d\lambda \quad (n = 1,2,\ldots). \tag{28.7}$$

---

*Adapted from J. Mikusiński [5], p. 414 ff.

Since $\hat{f}(\lambda_2) - \hat{f}(\lambda_1) = \{\hat{f}(\lambda_2,t) - \hat{f}(\lambda_1,t)\} \in C[0,\infty)$ and $(s-n\omega)^{-1} = \{e^{n\omega t}\}$, we see that the left hand side of (28.7) is in $C[0,\infty)$. Hence,

$$\left|\frac{\hat{f}(\lambda_2,t) - \hat{f}(\lambda_1,t)}{s - n\omega}\right| \leqq \int_{\lambda_1}^{\lambda_2} |\hat{f}(\lambda,t)|\,d\lambda \qquad (n = 1,2,\ldots). \tag{28.7'}$$

Thus, letting

$$g = \{g(t)\} = \left\{\frac{\hat{f}(\lambda_2,t) - \hat{f}(\lambda_1,t)}{s}\right\}$$

$$k = \{k(t)\} = \left\{\int_{\lambda_1}^{\lambda_2} |\hat{f}(\lambda,t)|\,d\lambda\right\},$$

we obtain, from (28.7)',

$$\left|\frac{g(t)}{s-n\omega}\right| = h\cdot\left|\frac{\hat{f}(\lambda_2,t) - \hat{f}(\lambda_1,t)}{s - n\omega}\right| \leqq h\{k(t)\};$$

that is,

$$\left|\int_0^t g(t-u)e^{n\omega u}\,du\right| \leqq \int_0^t k(u)\,du \qquad (n = 1,2,\ldots). \tag{28.8}$$

Therefore, by Theorem 20 (§21), we obtain the result

$$g(t-u) = 0 \quad \text{for all } u \text{ with } 0 \leqq u \leqq t,$$

i.e.,

$$g(u) = 0 \quad \text{for all } u \in [0,t].$$

Since $t > 0$ was chosen arbitrarily, we have proved that

$$\hat{f}(\lambda_2,t) - \hat{f}(\lambda_1,t) = 0 \quad \text{for all } t > 0, \text{ i.e., } \hat{f}(\lambda_2) = \hat{f}(\lambda_1).$$

Again, since $\lambda_1$ and $\lambda_2$ were arbitrary except for the restriction that $\lambda_1 < 0 < \lambda_2$, we must have the result that $f(\lambda)$ does not depend on $\lambda$. This proves that $f(\lambda) = f(0) = I$.

Hence $f'(\lambda) = 0$ and so, by (28.5), $f(\lambda) = 0$ (for all $\lambda$), contradicting $f(0) = I$.

PROPOSITION 26. Let both $w_1$ and $w_2 \in C/C$ be logarithmic hyperfunctions. Then any linear combination

$$(\alpha_1 w_1 + \alpha_2 w_2) \tag{28.9}$$

with real coefficients $\alpha_1$ and $\alpha_2$ is also a logarithmic hyperfunction. Moreover, we have

$$\exp(\lambda\alpha_1 w_1)\exp(\lambda\alpha_2 w_2) = \exp(\lambda(\alpha_1 w_1 + \alpha_2 w_2)) \quad (-\infty < \lambda < \infty). \tag{28.10}$$

PROOF: By applying (23.7) to

$$f(\lambda) = \exp(\lambda\alpha_1 w_1)\exp(\lambda\alpha_2 w_2) \tag{28.11}$$

we obtain

$$\begin{aligned} f'(\lambda) &= \alpha_1 w_1 \exp(\lambda\alpha_1 w_1)\exp(\lambda\alpha_2 w_2) \\ &\qquad + \exp(\lambda\alpha_1 w_1)\alpha_2 w_2 \exp(\lambda\alpha_2 w_2) \\ &= (\alpha_1 w_1 + \alpha_2 w_2) f(\lambda). \end{aligned}$$

Furthermore, it is true that $f(0) = I^2 = I$.

Therefore, by the uniqueness of the solution $\hat{f}(\lambda)$ of

$$\hat{f}'(\lambda) = (\alpha_1 w_1 + \alpha_2 w_2)\hat{f}(\lambda) \quad (-\infty < \lambda < \infty), \quad \hat{f}(0) = I,$$

we must have (28.10).

REMARK 28.1. As a Corollary of Proposition 26, we see that the operator $0 \in C/C$ is a logarithmic hyperfunction so that

$$f(\lambda) = I$$

is the unique solution of

$$f'(\lambda) = 0f(\lambda) = 0 \quad (-\infty < \lambda < \infty), \quad f(0) = I. \tag{28.12}$$

EXAMPLES OF THE LOGARITHMIC HYPERFUNCTIONS

EXAMPLE 28.1. For any $\{f(t)\} \in C$, $w = \{f(t)\}$ is a logarithmic hyperfunction.

PROOF: Let $f^o = I$, $f^1 = f$, $f^2 = f*f$, $f^3 = f*f^2, \ldots, f^n = f*f^{n-1}$. Then, as we already know in (25.4), we have, for any $T > 0$,

$$\begin{cases} |f^n(t)| \leqq M^n t^{n-1}/(n-1)! \quad (0 \leqq t \leqq T;\ n = 1,2,\ldots) \\ M_T = \sup\limits_{0 \leqq t \leqq T} |f(t)| = M < +\infty. \end{cases}$$

Hence

$$\exp(\lambda f) = I + \{f(\lambda,t)\} = \sum_{n=0}^{\infty} \frac{\lambda^n}{n!} f^n \in C_H \subseteqq C/C. \tag{28.13}$$

is defined as the generalized $\lim\limits_{k\to\infty} \Sigma_{n=0}^k \frac{\lambda^n}{n!} f^n \in C_H \subseteqq C/C$, and

$$\frac{d\,\exp(\lambda f)}{d\lambda\lambda} = \{\frac{\partial}{\partial\lambda} f(\lambda,t)\} = \sum_{n=1}^{\infty} \frac{\lambda^{n-1}}{(n-1)!} f^n$$
$$= f\cdot \sum_{n=0}^{\infty} \frac{\lambda^n}{n!} f^n = f\cdot\exp(\lambda f), \quad \exp(0\cdot f) = I. \tag{28.14}$$

As an application of the above example, we have

PROPOSITION 27. Let $\alpha$ be a positive real number. Then

$$w = s - \sqrt{s^2 + \alpha^2}$$

is a logarithmic hyperfunction and

$$\exp(\lambda(s - \sqrt{s^2 + \alpha^2})) = I - \left\{\frac{\lambda}{\sqrt{t^2+2\lambda t}} \alpha J_1(\alpha\sqrt{t^2 + 2\lambda t})\right\} \tag{28.15}$$

PROOF: By (16.7), we have

$$\sqrt{s^2 + \alpha^2} - s = \left\{\frac{\alpha}{t} J_1(\alpha t)\right\} \in C$$

and so, by (28.13),

$$\left\{\begin{aligned} &\exp(\lambda(s - \sqrt{s^2 + \alpha^2})) \\ &= I + \sum_{n=0}^{\infty} \frac{1}{(n+1)!} (-\lambda)^{n+1} (\sqrt{s^2 + \alpha^2} - s)^{n+1}. \end{aligned}\right. \tag{28.16}$$

On the other hand, we have, by (16.8)',

$$(\sqrt{s^2+\alpha^2} - s)^{n+1} = \left\{(-1)^n (n+1) \sum_{k=n}^{\infty} (-1)^k \frac{\alpha^{2+2k} t^{2k-n}}{2^{1+2k-n}(k-n)!(k+1)!}\right\}.$$

We substitute this in (28.16) and exchange the order of summation to obtain

$$\exp(\lambda(s - \sqrt{s^2 + \alpha^2}))$$
$$= I - \left\{\lambda \sum_{k=0}^{\infty} (-1)^k \frac{\alpha^{2+2k} t^k}{2^{1+2k} k!(k+1)!} \sum_{n=0}^{k} \frac{k!\lambda^n t^{k-n}}{2^{-n}(k-n)!n!}\right\}$$
$$= I - \left\{\lambda \sum_{k=0}^{\infty} (-1)^k \frac{\alpha^{2+2k} t^k}{2^{1+2k} k!(k+1)!} (t+2\lambda)^k\right\}$$
$$= I - \left\{\frac{\lambda}{\sqrt{t^2+2\lambda t}} \alpha J_1(\alpha\sqrt{t^2+2\lambda t})\right\} \quad \text{(by (15.3))}.$$

COROLLARY 1.

$$\frac{\exp(\lambda(s - \sqrt{s^2+\alpha^2}))}{\sqrt{s^2 + \alpha^2}} = \left\{J_0(\alpha\sqrt{t^2+2\lambda t})\right\} \tag{28.17}$$

PROOF: Differentiating (28.16) with respect to $\alpha$ and then dividing by $(-\alpha\lambda)$, we have

$$\frac{\exp(\lambda(s - \sqrt{s^2+\alpha^2}))}{\sqrt{s^2 + \alpha^2}} = \left\{\sum_{k=0}^{\infty} (-1)^k \frac{\alpha^{2k}t^k}{2^{2k}(k!)^2} (t + 2\lambda)^k\right\}.$$

Then we substitute the expansion of $J_0(t)$ ((15.3)) into the right hand side.

COROLLARY 2. Formulae (28.15) and (28.17) were obtained by making use of binomial expansions with respect to $\alpha$. Thus these formulae hold for $\lambda \geqq 0$ and for any complex number $\alpha \neq 0$. In particular, by substituting $i\alpha = \sqrt{-1}\,\alpha$ for $\alpha$, we obtain

$$\exp(\lambda(s - \sqrt{s^2 - \alpha^2})) = I - \left\{\frac{\lambda}{\sqrt{t^2+2\lambda t}}\, i\alpha J_1(i\alpha\sqrt{t^2 + 2\lambda t})\right\}, \tag{28.18}$$

$$\frac{\exp(\lambda(s - \sqrt{s^2 - \alpha^2}))}{\sqrt{s^2 - \alpha^2}} = \left\{J_0(i\alpha\sqrt{t^2 + 2\lambda t})\right\}. \tag{28.19}$$

Finally, multiplying (28.15), (28.17), (28.18) and (28.19) by $\exp(-\lambda s)$ and making use of (28.10) and of the shift effect of $\exp(-\lambda s)$, we obtain

COROLLARY 3.

$$\exp(-\lambda\sqrt{s^2+\alpha^2}) = e^{-\lambda s} - \begin{Bmatrix} 0, & 0 \leqq t < \lambda \\ \dfrac{\lambda}{\sqrt{t^2-\lambda^2}}\,\alpha J_1(\alpha\sqrt{t^2-\lambda^2}), & 0 \leqq \lambda < t \end{Bmatrix} \tag{28.20}$$

$$\frac{\exp(-\lambda\sqrt{s^2+\lambda^2})}{\sqrt{s^2 + \lambda^2}} = \begin{Bmatrix} 0, & 0 \leqq t < \lambda \\ J_0(\alpha\sqrt{t^2-\lambda^2}), & 0 \leqq \lambda < t \end{Bmatrix} \tag{28.21}$$

$$\exp(-\lambda\sqrt{s^2-\lambda^2}) = e^{-\lambda s} - \begin{Bmatrix} 0, & 0 \leqq t < \lambda \\ \dfrac{\lambda}{\sqrt{t^2-\lambda^2}}\, i\alpha J_1(i\alpha\sqrt{t^2-\lambda^2}), & 0 \leqq \lambda < t \end{Bmatrix}$$

(28.22)

$$\frac{\exp(-\lambda\sqrt{s^2-\alpha^2})}{\sqrt{s^2-\alpha^2}} = \begin{cases} 0, & 0 \leqq t < \lambda \\ J_0(i\alpha\sqrt{t^2-\lambda^2}), & 0 \leqq \lambda < t \end{cases} \tag{28.23}$$

EXERCISES FOR §28.

Let $w \in C/C$ be a logarithmic hyperfunction and let $f(\lambda)$ be a complex-valued continuous function defined for $-\infty < \lambda < \infty$. Prove that

$$z(\lambda) = z_0 \exp((\lambda-\lambda_0)w) + \int_{\lambda_0}^{\lambda} \exp((\lambda-\mu)w)f(\mu)d\mu \tag{28.24}$$

is a solution of

$$\frac{dz(\lambda)}{d\lambda} - wz(\lambda) = f(\lambda)I, \quad z(\lambda_0) = z_0. \tag{28.25}$$

# Part IV
# Applications to Partial Differential Equations

The purpose of this Part IV is to discuss mainly

*Wave Equation:* $Z_{\lambda\lambda}(\lambda,t) = \alpha^2 Z_{tt}(\lambda,t)$,

*Telegraph Equation:* $Z_{\lambda\lambda}(\lambda,t) = Z_{tt}(\lambda,t) + 2\kappa Z_t(\lambda,t) + \gamma Z(\lambda,t)$,

*Heat Equation:* $Z_{\lambda\lambda}(\lambda,t) = \alpha^2 Z_t(\lambda,t)$,

by converting these equations into hyperfunction equations.

The wave equation shall be discussed in the domain $\lambda_1 \leqq \lambda \leqq \lambda_2$, $t \geqq 0$. By making use of (5.6), we obtain

$$s^2 Z(\lambda,t) = Z_{tt}(\lambda,t) + Z_t(\lambda,0) + sZ(\lambda,0),$$

so that the initial condition with respect to $t$ will be

$$Z(\lambda,0) = \phi(\lambda), \quad Z_t(\lambda,0) = \Psi(\lambda).$$

Hence the hyperfunction equation for the solution $Z(\lambda) = \{Z(\lambda,t)\}$ of the wave equation is

$$\text{(i)} \quad Z''(\lambda) - \alpha^2 s^2 Z(\lambda) = -\alpha^2 \psi(\lambda) - \alpha^2 s\phi(\lambda).$$

It is to be remarked that the right hand side of the equation is a *linear function of* $s$.

The heat equation shall also be discussed in the domain $\lambda_1 \leqq \lambda \leqq \lambda_2$, $t \geqq 0$. Then, by (5.6), we obtain

$$sZ(\lambda) = Z_t(\lambda,t) + Z(\lambda,0),$$

and hence, putting $Z(\lambda,0) = \phi(\lambda)$, the hyperfunction equation for the solution $Z(\lambda) = \{Z(\lambda,t)\}$ of the heat equation is

(ii) $Z''(\lambda) - \alpha^2 sZ(\lambda) = -\alpha^2\phi(\lambda)$.

We are thus led to discuss the following problem. Let $w \in C/C$ be a *logarithmic hyperfunction*. Consider a hyperfunction-valued function $f(\lambda) = \{f(\lambda,t)\}$ which satisfies

(iii) $f''(\lambda) - wf(\lambda) = g(\lambda) \quad (\lambda_1 \leqq \lambda \leqq \lambda_2,\ 0 \leqq t)$,

where $g(\lambda)$ is a linear function of $w$, the coefficients of this linear function being numerical valued functions of $\lambda$.

The case $w = s$ is the wave equation (Chapter IX) and the case $w = s^{1/2}$ is the heat equation (XI). We shall discuss the telegraph equation in Chapter X.

REMARK. In the case of *the vibration of a string*, let $Z(\lambda,t)$ be the value of the vertical displacement of the point of the string (with the abscissa $\lambda$, at the instant $t$) from the equilibrium position of the string (= the $\lambda$ axis). Then the customary assumption that the partial derivative

$$\frac{\partial^2}{\partial\lambda^2} Z(\lambda,t)$$

is continuous at every point $\lambda$ of the string might be somewhat restrictive from the physical point of view. In this sense, our approach, using the generalized derivative $Z''(\lambda)$, would be easier to handle as the reader will see in due course in this Part IV (see, e.g., Remark 30.1 in the Section 30 of this book).

# Chapter IX
# One Dimensional Wave Equation

## §29. HYPERFUNCTION EQUATION OF THE FORM $f''(\lambda) - w^2 f(\lambda) = g(\lambda)$, $w \in C/C$

PROPOSITION 28. Let $w \in C/C$, $k_0 \in C/C$ and $k_1 \in C/C$. Further, let $w$ be a logarithmic hyperfunction and $w \neq 0$. Then the solution of the equation

$$f''(\lambda) - w^2 f(\lambda) = 0; \quad f(0) = k_0, \quad f'(0) = k_1, \quad (-\infty < \lambda < \infty) \tag{29.1}$$

is uniquely determined from the $\lambda$-*initial condition** $f(0) = k_0$ and $f'(0) = k_1$ as follows:

$$\begin{cases} f(\lambda) = 2^{-1}(k_0 + w^{-1}k_1)\exp(\lambda w) \\ \qquad + 2^{-1}(k_0 - w^{-1}k_1)\exp(-\lambda w) \quad (-\infty < \lambda < \infty). \end{cases} \tag{29.2}$$

PROOF: It is apparent that both

$$\exp(\lambda w) \quad \text{and} \quad \exp(-\lambda w)$$

are solutions of the equation

$$f''(\lambda) - w^2 f(\lambda) = 0, \tag{29.1$'$}$$

and any linear combination

$$f_1(\lambda) = a\exp(\lambda w) + b\exp(-\lambda w) \tag{29.3}$$

with coefficients $a$ and $b \in C/C$ satisfies (29.1)' by Proposition 19 (§23). We shall determine $a$ and $b$ so that $f_1(\lambda)$ satisfies the $\lambda$-initial condition; i.e.,

*This is different from the initial condition with respect to $t$.

$$f_1(0) = a + b = k_0, \quad f_1'(0) = aw - bw = k_1. \tag{29.4}$$

Hence

$$aw + bw = k_0 w, \quad aw - bw = k_1,$$

and so, by virtue of the condition $w \neq 0$, we obtain the exitence of $w^{-1} = \frac{I}{w}$ (see (18.6)) and

$$a = 2^{-1}(k_0 + w^{-1}k_1), \qquad b = 2^{-1}(k_0 - w^{-1}k_1). \tag{29.5}$$

Next we shall prove the uniqueness of the solution of (29.1). Thus, assume that there exists another solution $f_2(\lambda)$ of (29.1). Then

$$\hat{f}(\lambda) = f_1(\lambda) - f_2(\lambda)$$

is a solution of

$$\hat{f}''(\lambda) - w^2\hat{f}(\lambda) = 0, \quad \hat{f}(0) = \hat{f}'(0) = 0 \quad (-\infty < \lambda < \infty). \tag{29.6}$$

In order to show that $\hat{f}(\lambda)$ must vanish, we take

$$\check{f}(\lambda) = \hat{f}'(\lambda)\hat{f}(2\mu-\lambda) + \hat{f}(\lambda)\hat{f}'(2\mu-\lambda), \tag{29.7}$$

where $\mu$ denotes any real number. Then, by Proposition 19, we have

$$\begin{aligned}\check{f}'(\lambda) &= \hat{f}''(\lambda)\hat{f}(2\mu-\lambda) - \hat{f}'(\lambda)f'(2\mu-\lambda)\\ &\qquad + \hat{f}'(\lambda)\hat{f}'(2\mu-\lambda) - \hat{f}(\lambda)\hat{f}''(2\mu-\lambda)\\ &= w^2\hat{f}(\lambda)\hat{f}(2\mu-\lambda) - \hat{f}(\lambda)w^2\hat{f}(2\mu-\lambda)\\ &= 0.\end{aligned}$$

Therefore, by Proposition 19, we have the result that $\check{f}(\lambda)$ does not depend on $\lambda$. Hence, by (29.7) and $\hat{f}(0) = \hat{f}'(0) = 0$, we have

$$\check{f}(\lambda) = \check{f}(0) = \hat{f}'(0)\hat{f}(2\mu) + \hat{f}(0)\hat{f}'(2\mu) = 0 \quad \text{for all } \lambda.$$

Thus, by putting $\mu = \lambda$ in (29.7), we obtain

$$0 = \check{f}(\lambda) = \hat{f}'(\lambda)\hat{f}(\lambda) + \hat{f}(\lambda)\hat{f}'(\lambda) = (\hat{f}(\lambda)^2)'.$$

Hence, again by Proposition 19, $\hat{f}(\lambda)^2$ does not depend on $\lambda$, and so

$$0 = \hat{f}(\lambda)^2 = \hat{f}(0)^2 \quad \text{for all } \lambda.$$

This implies, by Theorem 16 (§17), that

$$\hat{f}(\lambda) = 0 \quad (-\infty < \lambda < \infty).$$

We have thus proved the uniqueness of the solution of (29.1).

PROPOSITION 29. Let $w \in C/C$ be a logarithmic hyperfunction and $w \neq 0$. Let, for $\lambda_1$ and $\lambda_2$ satisfying $\lambda_1 < 0 < \lambda_2$, $\phi(\lambda)$ and $\psi(\lambda)$ be complex-valued continuous functions defined for $\lambda \in [\lambda_1, \lambda_2]$. Then, for

$$g(\lambda) = -w\phi(\lambda) - \psi(\lambda), \tag{29.8}$$

$$\begin{cases} f_0(\lambda) = -(2w)^{-1} \displaystyle\int_{\lambda_1}^{\lambda} \exp((\mu-\lambda)w)g(\mu)d\mu \\ \qquad\qquad -(2w)^{-1} \displaystyle\int_{\lambda}^{\lambda_2} \exp((\lambda-\mu)w)g(\mu)d\mu \end{cases} \tag{29.9}$$

is a solution of the hyperfunction equation

$$f_0''(\lambda) - w^2 f_0(\lambda) = g(\lambda) \quad (\lambda_1 < \lambda < \lambda_2). \tag{29.10}$$

PROOF: We have, by calculating the generalized derivative of (29.9),

$$\begin{aligned} f_0'(\lambda) &= \frac{-I}{2w}\left\{g(\lambda) - \int_{\lambda_1}^{\lambda} w \exp((\mu-\lambda)w)g(\mu)d\mu \right. \\ &\qquad \left. - g(\lambda) + \int_{\lambda}^{\lambda_2} w \exp((\lambda-\mu)w)g(\mu)d\mu\right\} \\ &= \frac{-I}{2w}\left\{\int_{\lambda_1}^{\lambda} - w \exp((\mu-\lambda)w)g(\mu)d\mu \right. \\ &\qquad \left. + \int_{\lambda}^{\lambda_2} w \exp((\lambda-\mu)w)g(\mu)d\mu\right\}. \end{aligned}$$

Hence

$$\begin{aligned} f_0''(\lambda) &= \frac{-I}{2w}\left\{-wg(\lambda) + \int_{\lambda_1}^{\lambda} w^2 \exp((\mu-\lambda)w)g(\mu)d\mu \right. \\ &\qquad \left. - wg(\lambda) + \int_{\lambda}^{\lambda_2} w^2 \exp((\lambda-\mu)w)g(\mu)d\mu\right\} \\ &= g(\lambda) + w^2 f_0(\lambda). \end{aligned}$$

THEOREM 24. Let $w \in C/C$ be a logarithmic hyperfunction and $w \neq 0$. Let both $k_0$ and $k_1$ be $\in C/C$. Let, further,

$$g(\lambda) = -w\phi(\lambda) - \psi(\lambda).$$

Then the equation

$$\begin{cases} f''(\lambda) - w^2 f(\lambda) = g(\lambda) \\ f(0) = k_0, \quad f'(0) = k_1 \end{cases} \quad (\lambda_1 < \lambda < \lambda_2)^*, \tag{29.11}$$

has a uniquely determined solution given by

$$f(\lambda) = f_0(\lambda) + a\exp(-\lambda w) + b\exp(\lambda w), \tag{29.12}$$

where $f_0(\lambda)$ is given by (29.9) and

$$\begin{cases} a = \dfrac{k_0}{2} - \dfrac{k_1}{2w} - \dfrac{f_0(0)}{2} + \dfrac{f_0'(0)}{2w} \\ b = \dfrac{k_0}{2} + \dfrac{k_1}{2w} - \dfrac{f_0(0)}{2} - \dfrac{f_0'(0)}{2w} \end{cases} \tag{29.12)'$$

PROOF: By (29.10), it is not difficult to show that (29.12) is a solution of

$$f''(\lambda) - w^2 f(\lambda) = g(\lambda).$$

To show that (29.12)-(29.12)' is a solution of (29.1), we shall verify the $\lambda$-intial condition. Thus

$$f(0) = f_0(0) + \frac{k_0}{2} - \frac{k_1}{2w} - \frac{f_0(0)}{2} + \frac{f_0'(0)}{2w} + \frac{k_0}{2} + \frac{k_1}{2w} - \frac{f_0(0)}{2} - \frac{f_0'(0)}{2w} = k_0,$$

$$f'(0) = f_0'(0) - \frac{k_0 w}{2} + \frac{k_1}{2} + \frac{f_0(0)w}{2} - \frac{f_0'(0)}{2} + \frac{k_0 w}{2} + \frac{k_1}{2} - \frac{f_0(0)w}{2} - \frac{f_0'(0)}{2} = k_1.$$

Finally, the uniqueness of the solution is proved as follows. Assume that there exists another solution $\tilde{f}$ of (29.1), and consider $\hat{f}(\lambda) = f(\lambda) - \tilde{f}(\lambda)$, where $f(\lambda)$ is given by (29.12). Thus this $\hat{f}(\lambda)$ is a solution of

$$\hat{f}''(\lambda) - w^2\hat{f}(\lambda) = 0, \quad \hat{f}(0) = \hat{f}'(0) = 0. \tag{29.11)'$$

Hence, as in the proof of Proposition 28, we prove that $\hat{f}(\lambda) = 0$ for all $\lambda$.

REMARK 29.1. Formula (29.12) shows that the solution of (29.1) is given by the sum of a *particular solution* $f_0(\lambda)$ of the *inhomogeneous equation*

---
* We assume that $\lambda_1 < 0 < \lambda_2$.

$$f''(\lambda) - w^2 f(\lambda) = g \tag{29.13}$$

(without mentioning $\lambda$-initial condition) and a *general solution* $a \exp(-\lambda w) + b \exp(\lambda w)$ of the *homogeneous equation*

$$f''(\lambda) - w^2 f(\lambda) = 0. \qquad (29.1)'$$

We then determine $a \in C/C$ and $b \in C/C$ in such a way that

$$f(\lambda) = f_0(\lambda) + a \exp(-\lambda w) + b \exp(\lambda w)$$

satisfies the $\lambda$-initial condition

$$f(0) = k_0, \quad f'(0) = k_1.$$

Hence the difficult part of solving (29.1) lies in obtaining a particular solution $f_0(\lambda)$ of (29.13).

There are cases where $g(\lambda)$ is a polynomial in $\lambda$ and in which we can obtain $f_0(\lambda)$ directly.

AN EXAMPLE.

$$y''(\lambda) - s^2 y(\lambda) = s(\lambda^3 + 1).$$

We substitute

$$y_0(\lambda) = a_0 + a_1\lambda + a_2\lambda^2 + a_3\lambda^3 \quad (a_0, \ldots, a_3 \in C/C)$$

into the equation and obtain

$$(2a_2 + 6a_3\lambda) - s(a_0 + a_1\lambda + a_2\lambda^2 + a_3\lambda^3) = s(\lambda^3 + 1).$$

Equating coefficients of $\lambda^k$ $(k = 0, 1, 2, 3)$, we have

$$2a_2 - s^2 a_0 = s1 = I, \quad 6a_3 - s^2 a_1 = 0, \quad -s^2 a_2 = 0, \quad -s^2 a_3 = s;$$

that is,

$$a_2 = 0, \quad a_3 = \frac{-I}{s}, \quad a_0 = \frac{-I}{s^2}, \quad a_1 = \frac{6a_3}{s^2} = \frac{-6I}{s^3}.$$

Hence we have

$$y_0(\lambda) = \frac{-I}{s^2} + \frac{-6\lambda I}{s^3} + \frac{-\lambda^3 I}{s}$$

$$= \{-t - 3t^2\lambda - \lambda^3\}.$$

EXERCISES FOR §29. Obtain a particular solution in 1 and 2 below.

1. $y''(\lambda) - s^2 y(\lambda) = s\lambda^3$.
2. $y''(\lambda) - s^2 y(\lambda) = s\lambda^4$.

## §30. THE VIBRATION OF A STRING

A string of length $\lambda_0$ is streched along the horizontal $\lambda$-axis from point 0 to point $\lambda_0$. Let $z(\lambda,t)$ be the value of the vertical displacement of the point of the string with the abscissa $\lambda$ at the instant $t$.

Then, as stated in the beginning of this Part IV, we have

$$z_{\lambda\lambda}(\lambda,t) = \alpha^2 z_{tt}(\lambda,t), \tag{30.1}$$

where the positive constant $\alpha = (\rho/P)^{1/2}$ is given by the tension $P$ of the string and $\rho$ the mass per unit of length.

At $t = 0$, the string coincides with the $\lambda$-axis, and, moreover, its particles have no velocity, so that

$$z(\lambda,0) = 0 \quad \text{and} \quad z_t(\lambda,0) = 0 \quad (0 \leqq \lambda \leqq \lambda_0). \tag{30.2}$$

Hence equation (30.1) converts into the hyperfunction equation

$$z''(\lambda) = \alpha^2 s^2 \; z(\lambda). \tag{30.3}$$

Further, let the left end $\lambda = 0$ of the string move in a direction perpendicular to the $\lambda$-axis, with its motion defined by the function $v_1(t)$; and similarly let the right end $\lambda = \lambda_0$ of the string move perpendicular to the $\lambda$-axis according to $v_2(t)$. Thus

$$z(0,t) = v_1(t) \quad \text{and} \quad z(\lambda_0,t) = v_2(t) \quad (0 \leqq t < \infty). \tag{30.4}$$

Therefore, we wish to solve (30.3) under the conditions

$$z(0) = v_1 \in C \;\; (\subseteqq C/C) \quad \text{and} \quad z(\lambda_0) = v_2 \in C \;\; (\subseteqq C/C). \tag{30.4'}$$

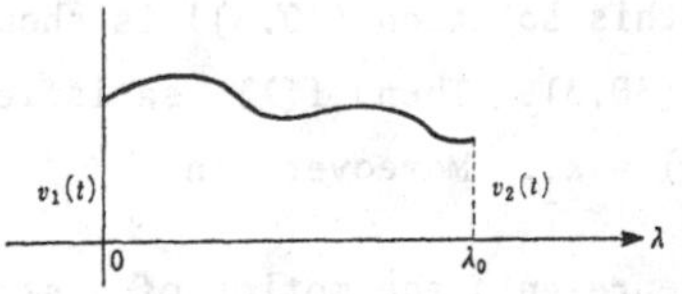

PROPOSITION 30. The (hyper)function $z(\lambda)$ satisfying equation (30.3) with condition (30.4)' can uniquely be represented as the sum

$$z(\lambda) = z_1(\lambda) + z_2(\lambda); \tag{30.5}$$

i.e.,

$$z(\lambda) = \frac{e^{-\alpha\lambda s} - e^{-\alpha(2\lambda_0-\lambda)s}}{I - e^{-2\alpha\lambda_0 s}} v_1 + \frac{e^{-\alpha(\lambda_0-\lambda)s} - e^{-\alpha(\lambda_0+\lambda)s}}{I - e^{-2\alpha\lambda_0 s}} v_2 \qquad (0 \leqq \lambda \leqq \lambda_0) \tag{30.5'}$$

so that $z_1(\lambda)$ and $z_2(\lambda)$ satisfy the same equation (30.3) and the conditions

$$\begin{cases} z_1(0) = v_1, & z_1(\lambda_0) = 0; \\ z_2(0) = 0, & z_2(\lambda_0) = v_2{}^* \end{cases}$$

respectively.

PROOF: The condition that the solution

$$z(\lambda) = a\exp(\alpha\lambda s) + b\exp(-\alpha\lambda s) \qquad (a, b \in C/C)$$

of (30.3) satisfies (30.4)' is given by

$$\begin{cases} z(0) = v_1 = a + b, \\ z(\lambda_0) = v_2 = ae^{\alpha\lambda_0 s} + be^{-\alpha\lambda_0 s} \end{cases} \tag{30.6}$$

This is solved by

$$\begin{cases} a = \dfrac{-e^{-2\alpha\lambda_0 s}v_1 + e^{-\alpha\lambda_0 s}v_2}{I - e^{-2\alpha\lambda_0 s}} \\[2ex] b = \dfrac{v_1 - e^{-\alpha\lambda_0 s}v_2}{I - e^{-2\alpha\lambda_0 s}} \end{cases} \tag{30.7}$$

Here the existence of the inverse $(I - e^{-2\alpha\lambda_0 s})^{-1}$ is given by Remark 25.1. We have thus proved that (30.5)' is a solution of (30.3)-(30.4)'.

The uniqueness of this solution (30.5)' is shown as follows. Let $f(\lambda)$ be a solution of (30.3). Then $f(\lambda)$ satisfies the $\lambda$-initial condition $f(0) = k_0$, $f'(0) = k_1$. Moreover, in

---

*The function $z_1(\lambda)$ represents the motion of a string whose right end is held fast, and the function $z_2(\lambda)$ represents the motion of a string whose left end is held fast.

$$x(\lambda) = a_1 \exp(\alpha\lambda s) + b_1 \exp(-\alpha\lambda s),$$

we can uniquely choose $a_1$ and $b_1$ from $C/C$ in such a way that

$$x(0) = a_1 + b_1 = k_0, \quad x'(0) = a_1\alpha s - b_1\alpha s = k_1,$$

since the determinant

$$\begin{vmatrix} 1 & 1 \\ \alpha s & -\alpha s \end{vmatrix} = -2\,\alpha s \neq 0.$$

Thus, by Proposition 28, $f(\lambda)$ must be of the form $x(\lambda)$. This fact insures that the (30.5)', obtained above, is the unique solution.

GRAPHICAL INTERPRETATION OF (30.5)'. How does the point of the string with abscissa $\lambda$ move when $t$ increases? We shall consider this for the solution $z_1(\lambda)$ in the case

$$\alpha = \frac{1}{2}, \quad \lambda_0 = 10 \quad \text{and} \quad v_1(t) = \begin{cases} t, & 0 \leqq t < 4 \\ 4, & 4 \leqq t < \infty \end{cases} \tag{30.8}$$

(Mikusiński [5], p. 203).

Then, for $\lambda = 2$, we have

$$z_1(2) = \frac{e^{-s} - e^{-9s}}{I - e^{-10s}}\, v_1. \tag{30.9}$$

(See the following three graphs.)

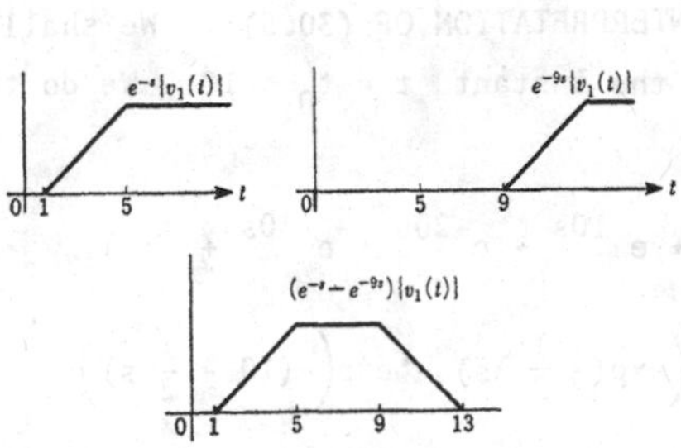

The graph of $e^{-\lambda s}\{v_1(t)\}$ is given by shifting the graph of $\{v_1(t)\}$ to the right by the distance 1. Similarly, the graph $e^{-9s}\{v_1(t)\}$ and the graph of $e^{-s}\{v_1(t)\} - e^{-9s}\{v_1(t)\}$ are given also.

Further, since

$$(I - e^{-10s})^{-1} = I + e^{-10s} + e^{-20s} + e^{-30s} + \cdots,$$

we have

$$z_1(2) = (I + e^{-10s} + e^{-20s} + e^{-30s} + \cdots)(e^{-s} - e^{-9s})v_1.$$

Hence the graph of $z_1(2)$ is given by the sum of the graphs which are obtained, respectively, by shifting the graph of

$$\{(e^{-s} - e^{-9s})v_1\}$$

to the right by the distance 0, the distance 10, the distance, 20, etc.

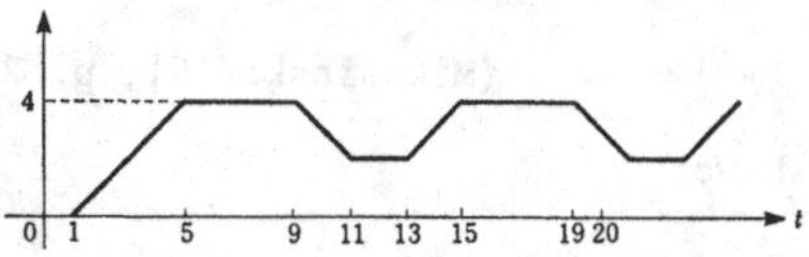


REMARK 30.1. $z_1(\lambda) = \{z_1(\lambda,t)\}$ is a hyperfunction solution of (30.3). Hence if $\{v_1(t)\}$ is not smooth, as in (30.8), then it is natural that the graph of $\{z_1(2,t)\}$ has many "corners".

GRAPHICAL INTERPRETATION OF (30.5)'. We shall consider the shape of the string at the instant $t = t_0 < 10$. We do this for the case (30.8). Then

$$z_1(\lambda) = (I + e^{-10s} + e^{-20s} + e^{-30s} + \cdots)$$

$$\times \left(\exp(-\tfrac{1}{2}\lambda s) - \exp\left(-(10 - \tfrac{\lambda}{2}s)\right)v_1\right.$$

$$= \sum_{n=0}^{\infty}\left\{\exp\left(-(10n + \tfrac{1}{2}\lambda)s\right) - \exp\left(-(10n + 10 - \tfrac{\lambda}{2})s\right)\right\}v_1$$

by making use of the addition theorem (28.4).

For this $z_1(\lambda)$, the terms in which the coefficients of $(-s)$ are larger than 10 vanish if $0 \leqq t < 10$ because of the property of the shift operator $e^{-\lambda s}$ given in Remark 22.2. Thus, to examine $z_1(\lambda)$ when $0 < t_0 < 10$ and $\lambda \leqq \lambda_0 = 10$, we have only to examine its term

$$\left[\exp(-\frac{1}{2}\lambda s) - \exp\left(-(10 - \frac{1}{2}\lambda)s\right)\right]\{v_1(t)\} = x_1(\lambda,t) - x_2(\lambda,t)$$

for $t = t_0$ and $\lambda \leqq 10$. By (30.8) and Remark 22.2, we have

$$\exp(-\frac{1}{2}\lambda s)\{v_1(t)\} = \begin{cases} 0 & (0 \leqq t < \frac{1}{2}\lambda) \\ t - \frac{1}{2}\lambda & (\frac{1}{2}\lambda \leqq t < \frac{1}{2}\lambda+4) \\ 4 & (\frac{1}{2}\lambda+4 \leqq t) \end{cases}$$

and

$$\exp -(10 - \frac{1}{2}\lambda)s\ \{v_1(t)\} = \begin{cases} 0 & (0 \leqq t < 10-\frac{1}{2}\lambda) \\ t-10+\frac{1}{2}\lambda & (10-\frac{1}{2}\lambda \leqq t < 14-\frac{\lambda}{2}) \\ 4 & (14-\frac{1}{2}\lambda \leqq t). \end{cases}$$

The graph of $x_1(\lambda,t_0) = \exp(-\frac{1}{2}\lambda s)\{v_1(t)\}_{t=t_0}$ is represented by a continuous line and the graph of $x_2(\lambda,t_0) = \exp(-(10 - \frac{\lambda}{2})s) \times \{v_1(t)\}_{t=t_0}$ is represented by a dotted line.

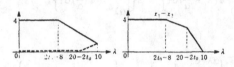

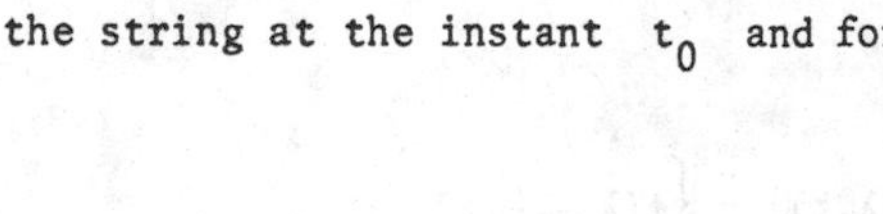


Subtracting, graphically, the dotted line from the continuous line, we obtain the shpae of the string at the instant $t_0$ and for $\lambda \leqq \lambda_0 = 10$.

## §31. D'ALEMBERT'S METHOD

We shall treat the vibration of a strong which is held fast at its two ends $\lambda = 0$ and $\lambda = \lambda_0$;

$$z(0,t) = 0, \quad z(\lambda_0,t) = 0 \quad (0 \leqq t < \infty), \tag{31.1}$$

with, furthermore, its initial shape $z(\lambda,0)$ at the instant $t = 0$ given by a real-valued continuous function $\phi(\lambda)$; i.e.,

$$z(\lambda,0) = \phi(\lambda) \quad (0 \leqq \lambda \leqq \lambda_0). \tag{31.2}$$

As was suggested by Jean le Rond D'Alembert (1717-1783), we shall make use of

PROPOSITION 31.* The generalized derivative with respect to $\lambda$ of

$$z_1(\lambda) = \{z_1(\lambda,t)\} = \left\{\phi(\lambda + \frac{t}{\alpha})\right\} \tag{31.3}$$

satisfies

$$z_1'(\lambda) = \alpha s z_1(\lambda) - \alpha[\phi(\lambda)], \quad \text{where} \quad [\phi(\lambda)] = s\{\phi(\lambda)\} = s\phi(\lambda). \tag{31.4}$$

PROOF: Since

$$z_1(\lambda) = s\left\{\int_0^t \phi(\lambda + \frac{\tau}{\alpha})d\tau\right\} = s\left\{\int_{\alpha\lambda}^{\alpha\lambda+t} \phi(\frac{\tau}{\alpha})d\tau\right\},$$

we have

$$z_1'(\lambda) = s\left\{\frac{\partial}{\partial\lambda}\int_{\alpha\lambda}^{\alpha\lambda+t} \phi(\frac{\tau}{\alpha})d\tau\right\} = \alpha s\left\{\phi(\lambda + \frac{t}{\alpha}) - \phi(\lambda)\right\}$$

$$= \alpha s z_1(\lambda) - \alpha[\phi(\lambda)].$$

COROLLARY 1. For

$$z_2(\lambda) = \{z_2(\lambda,t)\} = \left\{\phi(\lambda - \frac{t}{\alpha})\right\} \tag{31.3'}$$

we have

$$z_2'(\lambda) = -\alpha s z_2(\lambda) + \alpha[\phi(\lambda)]. \tag{31.4'}$$

COROLLARY 2. The function

$$z(\lambda) = \frac{1}{2}(z_1(\lambda) + z_2(\lambda)) = \frac{1}{2}\left\{\phi(\lambda + \frac{t}{\alpha}) + \phi(\lambda - \frac{t}{\alpha})\right\} \tag{31.5}$$

*J. Mikusiński [5], p. 214.

satisfies the hyperfunction equation

$$z''(\lambda) = \alpha^2 s^2 z(\lambda) - \alpha^2 s[\phi(\lambda)] \qquad (0 \leqq \lambda \leqq \lambda_1), \tag{31.6}$$

which corresponds to the string vibration equation and initial conditions as follows:

$$\begin{cases} z_{\lambda\lambda}(\lambda,t) = \alpha^2 z_{tt}(\lambda,t) & (0 \leqq t < \infty), \\ z(\lambda,0) = \phi(\lambda), \quad z_t(\lambda,0) = 0 & (0 \leqq \lambda \leqq \lambda_0). \end{cases}$$

PROOF: Since

$$z'(\lambda) = \frac{1}{2}(z_1'(\lambda) + z_2'(\lambda)) = \frac{1}{2}\alpha s(z_1(\lambda) - z_2(\lambda)),$$

we have

$$\begin{aligned} z''(\lambda) &= \frac{1}{2}\alpha s(z_1'(\lambda) - z_2'(\lambda)) \\ &= \frac{1}{2}\alpha^2 s^2(z_1(\lambda) + z_2(\lambda)) - \frac{1}{2} 2\alpha^2 s[\phi(\lambda)] \\ &= \alpha^2 s^2 z(\lambda) - \alpha^2 s[\phi(\lambda)]. \end{aligned}$$

Moreover, $z(\lambda,0) = \phi(\lambda)$ is clear and, if $z(\lambda)$ is twice continuously differentiable with respect to $t$, then, by (5.6),

$$z_{tt}(\lambda,t) = s^2 z(\lambda,t) - s[z(\lambda,0)] - [z_t(\lambda,0)].$$

REMARK 31.1. In order that

$$z(\lambda) = \frac{1}{2}\left\{\phi(\lambda + \frac{t}{\alpha}) + \phi(\lambda - \frac{t}{\alpha})\right\}$$

satisfies the $\lambda$-boundary condition (31.1), we must have

$$\begin{cases} \frac{1}{2}\left(\phi(\frac{t}{\alpha}) + \phi(\frac{-t}{\alpha})\right) = 0, \\ \frac{1}{2}\left(\phi(\lambda_0 + \frac{t}{\alpha}) + \phi(\lambda_0 - \frac{t}{\alpha})\right) = 0 \end{cases} \qquad (0 \leqq t < +\infty) \tag{31.7}$$

This condition (31.7) is surely satisfied if

$$-\phi(-\lambda) = \phi(\lambda) = \phi(\lambda + 2\lambda_0). \tag{31.8}$$

Thus if $\phi(\lambda)$ is an odd function defined for all real $\lambda$ and, moreover, is periodic with period $2\lambda_0 > 0$, then $\phi(\lambda)$ satisfies (31.8). Such is the case, for example, if $\phi(\lambda) = \sin(\pi\lambda/\lambda_0)$.

Therefore, if we take $\phi(\lambda)$ as above, then $z(\lambda)$ given by (31.5) is a hyperfunction solution of (31.6) satisfying (31.1)-(31.2).

REMARK 31.2. Any solution $\hat{z}(\lambda)$ of (31.6) satisfying the $\lambda$-boundary condition

$$\hat{z}(0) = 0, \quad \hat{z}(\lambda_0) = 0,$$

must be given by

$$z(\lambda) = \frac{1}{2}\left\{\phi(\lambda + \frac{t}{\alpha}) + \phi(\lambda - \frac{t}{\alpha})\right\}$$

with $\phi(\lambda)$ as above.

PROOF: Let $\hat{z}(\lambda)$ be such a solution and let

$$\tilde{z}(\lambda) = z(\lambda) - \hat{z}(\lambda).$$

Then we have

$$z''(\lambda) = \alpha^2 s^2 \tilde{z}(\lambda), \quad \tilde{z}(0) = 0, \quad \tilde{z}(\lambda_0) = 0. \tag{31.9}$$

Therefore, as in the proof of Proposition 30, the solution of (31.9) is uniquely determined so that $\tilde{z}(\lambda)$ must vanish, where $\hat{z}(\lambda) = z(\lambda)$.

THE STRING VIBRATION WITH GIVEN INITIAL VELOCITY. We shall determine the motion of a string under the *initial condition*

$$z(\lambda,0) = 0, \quad z_t(\lambda,0) = \psi(\lambda) \quad (0 \leqq \lambda \leqq \lambda_0), \tag{31.10}$$

and the *boundary condition*

$$z(0,t) = 0, \quad z(\lambda_0,t) = 0 \quad (0 \leqq t < \infty), \tag{31.11}$$

assuming that no external force acts upon the string. The corresponding hyperfunction equation is, by (5.6),

$$z''(\lambda) = \alpha^2 s^2 z(\lambda) - \alpha^2[\psi(\lambda)]. \tag{31.12}$$

Hence if $z(\lambda) = \{z(\lambda,t)\}$ satisfies (31.10), (31.11) and (31.12), then

$$y(\lambda) = sz(\lambda)$$

satisfies

$$\begin{cases} y''(\lambda) = \alpha^2 s^2 y(\lambda) - \alpha^2 s[\psi(\lambda)], \\ y(0) = y(\lambda_0) = 0. \end{cases} \tag{31.13}$$

Thus, as in the case of (31.5), we obtain

$$y(\lambda) = \frac{1}{2}\left\{\psi(\lambda + \frac{t}{\alpha}) + \psi(\lambda - \frac{t}{\alpha})\right\};$$

so that

$$hy(\lambda) = z(\lambda) = \{z(\lambda,t)\} = \frac{1}{2}\left\{\int_0^t \left(\psi(\lambda + \frac{\tau}{\alpha}) + \psi(\lambda - \frac{\tau}{\alpha})\right)d\tau\right\} \quad (31.14)$$

would be a candidate for the unique hyperfunction solution of (31.12) under conditions (31.10) and (31.11). As in the case of (31.7), we see that (31.14) satisfies (31.10) and (31.11) if $\psi(\lambda)$ satisfies

$$-\psi(-\lambda) = \psi(\lambda) = \psi(\lambda + 2\lambda_0). \quad (31.8)'$$

D'ALEMBERT'S WAVE FORMULA. We have thus obtained the following result. If $\phi(\lambda)$ and $\psi(\lambda)$ are respectively twice and once continuously differentiable, then, under the condition (31.8)-(31.8)', the so-called D'Alembert's formula

$$z(\lambda,t) = \frac{1}{2}\left\{\phi(\lambda - \frac{t}{\alpha}) + \phi(\lambda + \frac{t}{\alpha}) + \alpha\int_{\lambda-t/\alpha}^{\lambda+t/\alpha} \psi(u)\,du\right\}$$

gives the solution of

$$z_{\lambda\lambda}(\lambda,t) = \alpha^2 z_{tt}(\lambda,t); \quad z(\lambda,0) = \phi(\lambda), \quad z_t(\lambda,0) = \psi(\lambda).$$

EXAMPLE 31.1. If $\phi(\lambda) = \sin\frac{\pi}{\lambda_0}\lambda$, the function

$$\begin{aligned} z(\lambda,t) &= \frac{1}{2}\left\{\sin\left(\frac{\pi}{\lambda_0}(\lambda - \frac{t}{\alpha})\right) + \sin\left(\frac{\pi}{\lambda_0}(\lambda + \frac{t}{\alpha})\right)\right\} \\ &= \cos\left(\frac{\pi}{\lambda_0}\frac{t}{\alpha}\right)\sin\left(\frac{\pi}{\lambda_0}\lambda\right) \end{aligned}$$

is the solution of

$$\begin{cases} z_{\lambda\lambda}(\lambda,t) = \alpha^2 z_{tt}(\lambda,t), \\ z(0,t) = 0, \quad z(\lambda_0,t) = 0, \quad z(\lambda,0) = \phi(\lambda), \quad z_t(\lambda,0) = 0. \end{cases}$$

EXAMPLE 31.2. If $\psi(\lambda) = \sin\frac{\pi}{\lambda_0}\lambda$, the function

$$\begin{aligned} z(\lambda,t) &= \frac{\alpha}{2}\int_0^t \left(\psi(\lambda + \frac{\tau}{\alpha}) + \psi(\lambda - \frac{\tau}{\alpha})\right)d\tau \\ &= \sin\frac{\pi\lambda}{\lambda_0}\int_0^t \cos\left(\frac{\pi}{\lambda_0}\frac{\tau}{\alpha}\right)d\tau = \frac{\alpha\lambda_0}{\pi}\sin\frac{\pi\lambda}{\lambda_0}\sin\left(\frac{\pi}{\lambda_0}\frac{t}{\alpha}\right) \end{aligned}$$

is the solution of

$$\begin{cases} z_{\lambda\lambda}(\lambda,t) = \alpha^2 z_{tt}(\lambda,t), \\ z(0,t) = 0, \quad z(\lambda_0,t) = 0, \quad z(\lambda,0) = 0, \quad z_t(\lambda,0) = \psi(\lambda). \end{cases}$$

## §32. THE VIBRATION OF AN INFINITELY LONG STRING

Consider the uniquely determined hyperfunction solution (see (30.5)')

$$z_1(\lambda) = \frac{e^{-\alpha\lambda s} - e^{-\alpha(2\lambda_0-\lambda)s}}{I - e^{-2\alpha\lambda_0 s}} v_1 \qquad (\alpha > 0) \tag{32.1}$$

of the equation

$$z_1''(\lambda) = \alpha^2 s^2 z_1(\lambda) \tag{32.2}$$

satisfying

$$\begin{cases} z_1(\lambda,0) = 0, \quad (z_1)_t(\lambda,0) = 0 \qquad (0 \leqq \lambda \leqq \lambda_0), \\ z_1(0,t) = \{v_1(t)\}, \quad z_1(\lambda_0,t) = 0 \qquad (0 \leqq t). \end{cases} \tag{32.3}$$

We shall examine $z_1(\lambda)$ for $0 \leqq t < \alpha\lambda_0$. Consider the decomposition of $z_1(\lambda)$:

$$\begin{cases} z_1(\lambda) = e^{-\alpha\lambda s} v_1 + r(\lambda) v_1, \\ r(\lambda) v_1 = \dfrac{e^{-\alpha(2\lambda_0+\lambda)s} v_1 - e^{-\alpha(2\lambda_0-\lambda)s} v_1}{I - e^{-2\alpha\lambda_0 s}}. \end{cases}$$

Then, by the property of the shift operator, we see that

$$\text{both} \quad e^{-\alpha(2\lambda_0+\lambda)s}\{v_1(t)\} \quad \text{and} \quad e^{-\alpha(2\lambda_0-\lambda)s}\{v_1(t)\}$$

vanish when

$$0 \leqq \lambda \leqq \lambda_0, \qquad 0 \leqq t < \alpha\lambda_0. \tag{32.4}$$

In fact, the conditions that

$$\text{both} \quad \alpha(2\lambda_0+\lambda) > t \quad \text{and} \quad \alpha(2\lambda_0-\lambda) > t$$

hold when (32.4) is satisfied.

Thus, if the string is *very long*, i.e., if $\lambda_0$ is *very large*, we may put $r(\lambda)\{v_1(t)\} = 0$ so that we have

$$z_1(\lambda) = e^{-\alpha\lambda s}\{v_1(t)\} = \begin{cases} 0, & 0 \leqq t < \alpha\lambda \\ v_1(t-\alpha\lambda), & 0 \leqq \alpha\lambda < t. \end{cases} \tag{32.5}$$

Therefore, it can be seen that at the instant $t$ the front of the wave moving from the initial point $\lambda = 0$ of the string at $t = 0$ reaches the point $\lambda = t/\alpha$. Thus the wave moves with the velocity

$$1/\alpha = \sqrt{P/\rho}.$$

Such is the motion of a string which is said — though not quite correctly — to be *infinitely long*.

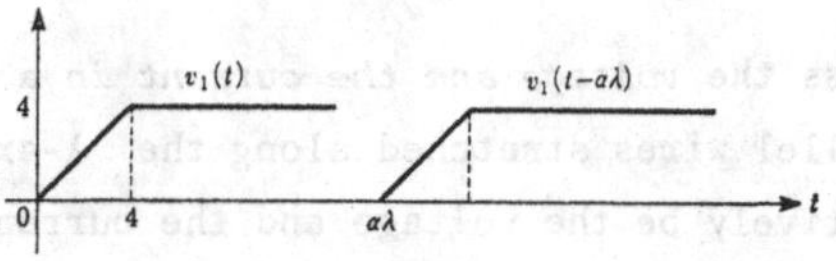


The above graph describes the case of

$$v_1(t) = \begin{cases} t, & 0 \leqq t < 4 \\ 4, & 4 \leqq t < \infty. \end{cases}$$

# Chapter X
# Telegraph Equation

## §33. THE HYPERFUNCTION EQUATION OF THE TELEGRAPH EQUATION

We shall discuss the *voltage and the current in a long cable* consisting of two parallel wires stretched along the $\lambda$-axis. Let $U(\lambda,t)$ and $I(\lambda,t)$ respectively be the voltage and the current at the point of the cable with coordinate $\lambda$ at the instant $t$. Then the following relations hold between the functions $U$ and $I$:

$$U_\lambda = -LI_t - RI, \quad I_\lambda = -CU_t - GU, \tag{33.1}$$

where $R$ denotes *resistance*, $L$ *self-inductance*, $G$ *leak-conductance* and $C$ *capacitance*; these quantities are measured per unit length of the cable.

Suppose that at the initial instant $t = 0$, no current flows through the cable and the voltage is equal to zero everywhere:

$$U(\lambda,0) = 0, \quad I(\lambda,0) = 0. \tag{33.2}$$

Hence

$$U_t(\lambda) = sU(\lambda), \quad I_t(\lambda) = sI(\lambda)$$

and so, by (33.1),

$$\begin{cases} U'(\lambda) = -LsI(\lambda) - RI(\lambda) = -(Ls + R)I(\lambda) \\ I'(\lambda) = -CsU(\lambda) - GU(\lambda) = -(Cs + G)U(\lambda). \end{cases} \tag{33.3}$$

This implies

$$U''(\lambda) = -(Ls+R)I'(\lambda) = (Ls+R)(Cs+G)U(\lambda).$$

We have thus obtained the equation for the voltage $U(\lambda)$:

$$U''(\lambda) = (LCs^2 + (RC+LG)s + RG)U(\lambda). \tag{33.4}$$

Similarly, we obtain the equation for the current $I(\lambda)$:

$$I''(\lambda) = (Cs+G)(Ls+R)I(\lambda) = (LCs^2 + (RC+LG)s + RG)I(\lambda). \qquad (33.4)'$$

These two equations (33.4) and (33.4)' are of the same form, and we call them *telegraph equations*.

REMARK 33.1. The telegraph equation reduces to the wave equation if $LC = \alpha^2$, $(RC + LG) = 0$ and $RG = 0$. Similarly, the telegraph equation reduces to the heat equation if $LC = 0$, $(RC + LG) = \alpha^2$ and $RG = 0$. We shall give, in the following sections, the concrete representation of the hyperfunction solution of the telegraph equation by classifying the equation according to the relationship among the parameters $L$, $C$, $G$ and $R$.

## §34. A CABLE WITH INFINITELY SMALL LOSS

Conductance never occurs without losses. They are due to the resistance $R$ and the leak-conductance $G$ of the current. However, if the quantities $R$ and $G$ are very small, self-inductance $L$ and capacity $C$ being large, we may assume approximately that

$$R = 0 \quad \text{and} \quad G = 0 \quad (\text{furthermore } L > 0 \text{ and } C > 0).$$

We shall call such a cable as *with infinitely small loss* or rather simply as *a cable without loss*.

The telegraph equation is then reduced to

$$U''(\lambda) = LCs^2U(\lambda) \quad (L > 0,\ C > 0). \qquad (34.1)$$

We further assume that the cable is *considerably long,* and that at the left end $\lambda = 0$ of the cable an electromotive force given by $E = \{E(t)\}$ is applied.

Thus we are confronted with (34.1) subject to the $\lambda$-boundary condition

$$U(0) = E = \{E(t)\}, \quad U(\lambda_0) = 0, \qquad (34.2)$$

where $\lambda_0 > 0$ is *considerably large*.

Therefore, as in the case of $z_i(\lambda)$ of §32, we can take

$$U(\lambda) = \exp(-\sqrt{LC}\,\lambda s)\{E(t)\} \qquad (34.3)$$

and so

$$U(\lambda,t) = \begin{cases} 0, & 0 \leqq t < \sqrt{LC}\,\lambda \\ E(t - \sqrt{LC}\lambda), & 0 \leqq \sqrt{LC}\lambda < t \end{cases}. \qquad (34.4)$$

Thus the wave of voltage moves with the velocity $1/\sqrt{LC}\lambda$. Similarly we have, by making use of (33.3), (34.3), and $R = 0$,

$$I(\lambda) = -\frac{1}{Ls}U'(\lambda) = \sqrt{C/L}\exp(-\sqrt{LC}\lambda s)\{E(t)\}; \tag{34.5}$$

that is,

$$I(\lambda,t) = \begin{cases} 0, & 0 \leqq t < \sqrt{LC}\,\lambda \\ \sqrt{C/L}\,E(t - \sqrt{LC}\lambda), & 0 \leqq \sqrt{LC}\,\lambda < t. \end{cases} \tag{34.6}$$

## §35. CONDUCTANCE WITHOUT DEFORMATION

It is said that O. Heaviside was the first to point out that in the more general case

$$\frac{R}{L} = \frac{G}{C} \quad (L > 0,\ C > 0) \tag{35.1}$$

both the voltage and the current are transferred in a way similar to the case of the preceding section, the only difference being the fact that the wave amplitudes are exponentially damped as the waves move forward.

Indeed, from (35.1), the telegraph equation

$$U''(\lambda) - (Ls+R)(Cs+G)U(\lambda) = 0$$

is written, by putting $L = \gamma C$ and $R = \gamma G$ with $\gamma > 0$, as

$$U''(\lambda) - \gamma(Cs+G)^2U(\lambda) = 0;$$

that is,

$$U''(\lambda) - (\alpha s+\beta)^2U(\lambda) = 0 \quad (\alpha = \sqrt{LC},\ \ \beta = \sqrt{RG}). \tag{35.2}$$

As in the preceding section, we assume

$$U(0) = E = \{E(t)\}, \quad U(\lambda_0) = 0 \tag{35.3}$$

and that the cable is considerably long. Thus we may prove as in the case of $z_1(\lambda)$ of §32, that

$$U(\lambda) = \{U(\lambda,t)\} = \exp(-(\alpha s+\beta)\lambda)\{E(t)\}. \tag{35.4}$$

PROOF: The condition that the general solution of (35.2),

$$U(\lambda) = ae^{(\alpha s+\beta)\lambda} + be^{-(\alpha s+\beta)\lambda} \quad (a,b \in C/C),$$

satisfies the $\lambda$-boundary condition (35.3) is given by

$$a+b = E = \{E(t)\}, \quad ae^{(\alpha s+\beta)\lambda_0} + be^{-(\alpha s+\beta)\lambda_0} = 0;$$

that is, by

$$a = \frac{-e^{-2(\alpha s+\beta)\lambda_0}\cdot E}{I-e^{-2(\alpha s+\beta)\lambda_0}}, \quad b = \frac{E}{I-e^{-2(\alpha s+\beta)\lambda_0}} .$$

Hence

$$U(\lambda) = \frac{e^{-(\alpha s+\beta)\lambda}\cdot E - e^{-(\alpha s+\beta)(2\lambda_0-\lambda)}\cdot E}{I - e^{-2(\alpha s+\beta)\lambda_0}}$$

$$= e^{-(s+\beta)\lambda}\cdot E + r(\lambda)\cdot E, \tag{35.5}$$

$$r(\lambda)E = \frac{e^{-(\alpha s+\beta)(2\lambda_0+\lambda)}\cdot E - e^{-(\alpha s+\beta)(2\lambda_0-\lambda)}\cdot E}{I - e^{-2(\alpha s+\beta)\lambda_0}} .$$

In the numerator of $r(\lambda)E$, both the two terms

$$e^{-\beta(2\lambda_0+\lambda)}\cdot e^{-\alpha(2\lambda_0+\lambda)s}\{E(t)\},$$

$$e^{-\beta(2\lambda_0-\lambda)}\cdot e^{-\alpha(2\lambda_0-\lambda)s}\{E(t)\}$$

are zero if

$$0 \leqq \lambda \leqq \lambda_0, \quad 0 \leqq t < \alpha\lambda_0 \tag{35.6}$$

because of the presence of shift operators

$$e^{-\alpha(2\lambda_0+\lambda)s} \text{ and } e^{-\alpha(2\lambda_0-\lambda)s} .$$

In fact, (35.6) implies

$$\alpha(2\lambda_0+\lambda) > t, \quad \alpha(2\lambda_0-\lambda) > t.$$

Thus when $\lambda_0 \to \infty$ we obtain $r(\lambda)E = 0$. Hence we have proved (35.4) for an infinitely long cable satisfying (35.1) and (35.3).

As a Corollary of (35.4), we have, by making use of $\alpha = \sqrt{LC}$, $\beta = \sqrt{RG}$ and $\exp(-(\alpha s+\beta))\lambda = e^{-\beta\lambda}e^{-\alpha\lambda s}$,

$$U(\lambda,t) = \begin{cases} 0, & 0 \leqq t < \sqrt{LC}\,\lambda \\ e^{-\lambda\sqrt{RG}}\,E(t-\sqrt{LC}\lambda), & 0 \leqq \sqrt{LC}\lambda < t \end{cases} . \tag{35.7}$$

Hence the amplitude $e^{-\lambda\sqrt{RG}}$ of the wave *decays exponentially* as $\lambda \to \infty$.

Similarly, by virtue of (33.3) we obtain

$$I(\lambda) = -\frac{U'(\lambda)}{Ls+R} = \frac{\alpha s+\beta}{Ls+R}\, e^{-(\alpha s+\beta)\lambda}\{E(t)\}$$

$$= \sqrt{C/L}\, e^{-(\alpha s+\beta)\lambda}\{E(t)\}.$$

Here we have made use of (35.1) to the effect that

$$(Ls+R)\sqrt{C/L} = \sqrt{LC}s + \sqrt{RG} = \alpha s + \beta.$$

Hence we have

$$I(\lambda,t) = \begin{cases} 0, & 0 \leqq t < \sqrt{LC}\,\lambda \\ \sqrt{C/L}\, e^{-\sqrt{RG}\lambda} E(t - \sqrt{LC}\lambda), & 0 \leqq \sqrt{CL}\lambda < t \end{cases}. \tag{35.8}$$

Thus, for $I(\lambda,t)$ as well, the amplitude $\sqrt{C/L}\, e^{-\sqrt{RG}\lambda}$ of the wave *decays exponentially* as $\lambda \to \infty$.

REMARK 35.1. The uniqueness of the solution given in (35.5) is proved as in the case of the solution given in (30.5)'.

## §36. THE THOMSON* CABLE

This is the case

$$L = G = 0 \quad (R > 0, \quad C > 0).$$

In this case, under the assumption of *considerably long cable,* Thomson calculated, in the middle of the 19th century, the propagation of current through a cable on the sea bottom across the Atlantic Ocean.

The telegraph equation is, when $L = G = 0$, of the form

$$U''(\lambda) = RCsU(\lambda). \tag{36.1}$$

Let the initial voltage at $t = 0$ be zero:

$$U(\lambda,0) = 0 \qquad (0 \leqq \lambda \leqq \lambda_0). \tag{36.2}$$

Here $\lambda_0$ is the length of the cable, which is sufficiently long. Let the electromotive force $E = \{E(t)\}$ applied at the initial point of the cable be constant in time:

$$U(0,t) = E = \{E_0\}, \quad U(\lambda_0,t) = 0 \quad (0 \leqq t < \infty). \tag{36.3}$$

Thus putting $\sqrt{RC} = \alpha > 0$, we have to solve the equation

$$U''(\lambda) = \alpha^2 sU(\lambda) \qquad (\alpha = \sqrt{RC} > 0) \tag{36.4}$$

under conditions (36.2) and (36.3).

*William Thomson (later Lord Kelvin, 1824-1907).

PROPOSITION 32. The hyperfunction solution of (36.4) combined with (36.2)-(36.3) is uniquely determined and is given by

$$U(\lambda) = \frac{(e^{-\alpha\lambda\sqrt{s}} - e^{-\alpha(2\lambda_0-\lambda)\sqrt{s}})E}{1 - e^{-2\alpha\lambda_0\sqrt{s}}}. \tag{36.5}$$

PROOF: For $a \in C/C$ and $b \in C/C$,

$$U(\lambda) = ae^{-\alpha\lambda\sqrt{s}} + be^{\alpha\lambda\sqrt{s}} \tag{36.6}$$

is the general solution of (36.4). We shall determine $a$ and $b$ so that (36.6) satisfies (36.3):

$$\begin{cases} U(0) = U(0,t) = a + b = E, \\ U(\lambda_0) = U(\lambda_0,t) = ae^{-\alpha\lambda_0\sqrt{s}} + be^{\alpha\lambda_0\sqrt{s}} = 0. \end{cases} \tag{36.7}$$

By solving (36.7), we obtain

$$\begin{cases} a = \dfrac{E}{I - e^{-2\alpha\lambda_0\sqrt{s}}}, \\ b = \dfrac{-e^{-2\alpha\lambda_0\sqrt{s}}E}{I-e^{-2\alpha\lambda_0\sqrt{s}}}. \end{cases} \tag{36.8}$$

Therefore, (36.5) is proved.

The uniqueness of the solution $U(\lambda)$ is proved as follows.

Let $F(\lambda)$ be a solution of (36.1). Then $F(\lambda)$ satisfies the $\lambda$-initial condition $F(0) = k_0$, $F'(0) = k_1$. Moreover, in

$$x(\lambda) = a_1e^{-\alpha\lambda\sqrt{s}} + b_1e^{\alpha\lambda\sqrt{s}},$$

we can uniquely choose $a_1$ and $b_1$ from $C/C$ in such a way that

$$x(0) = a_1 + b_1 = k_0, \quad x'(0) = -a_1\alpha\sqrt{s} + b_1\alpha\sqrt{s} = k_1, \tag{36.9}$$

since the determinant

$$\begin{vmatrix} 1 & 1 \\ -\alpha\sqrt{s} & \alpha\sqrt{s} \end{vmatrix} = 2\alpha\sqrt{s}.$$

is nonzero. Thus, by Theorem 24, $F(\lambda)$ must be of the form $x(\lambda)$. This fact assures that (36.5), obtained above, is the unique solution.

PROPOSITION 33. The unique solution $U(\lambda)$ given by (36.5) is expanded as follows.

$$\begin{cases} U(\lambda) = Ee^{-\alpha\lambda\sqrt{s}} + r(\lambda), \\ r(\lambda) = E\left(\sum_{\nu=1}^{\infty} (e^{-\alpha(2\nu\lambda_0+\lambda)\sqrt{s}} - e^{-\alpha(2\nu\lambda_0-\lambda)\sqrt{s}})\right). \end{cases} \tag{36.10}$$

PROOF: Expanding similarly as in (25.13), we have

$$\left(I - e^{-2\alpha\lambda_0\sqrt{s}}\right)^{-1} = I + \sum_{\nu=1}^{\infty} e^{-2\alpha\beta\lambda_0\sqrt{s}}$$

Hence, from (36.5), we obtain (36.10).

In order to estimate $r(\lambda)$ when $\lambda_0 \to \infty$, we shall make use of

PROPOSITION 34.* If $\lambda > 0$ and $0 < t < \infty$, we have

$$f(\lambda,t) = \frac{\lambda}{2\sqrt{\pi t^3}} \exp\left(-\frac{\lambda^2}{4t}\right) \leqq f\left(\lambda, \frac{\lambda^2}{6}\right) = \frac{3}{\lambda^2}\sqrt{\frac{6}{\pi e^3}}. \tag{36.11}$$

PROOF: Since

$$\frac{\partial}{\partial t} f(\lambda,t) = (\lambda^2-6t)\frac{\lambda}{8\sqrt{\pi t^7}} \exp\left(-\frac{\lambda^2}{4t}\right), \tag{36.12}$$

the function $f(\lambda,t)$ has a maximum at $t = \lambda^2/6$. Moreover, by $\lim_{t\downarrow 0} f(\lambda,t) = 0$ and $\lim_{t\to\infty} f(\lambda,t) = 0$, we see that $f(\lambda,t)$ attains, for $\lambda > 0$, its largest value $f(\lambda, \frac{\lambda^2}{6})$ at $t = \lambda^2/6$.

REMARK 36.1. (36.11) may be written as

$$e^{-\lambda\sqrt{s}} \leqq \left[\frac{3}{\lambda^2}\sqrt{\frac{6}{\pi e^3}}\right] h \tag{36.11$'$}$$

in the sense of hyperfunctions; that is, the left hand term is the hyperfunction represented by

$$\left\{\frac{\lambda}{2\sqrt{\pi t^3}} \exp\left(-\frac{\lambda^2}{4t}\right)\right\}$$

and the right hand term is the hyperfunction represented by

$$\left\{\frac{3}{\lambda^2}\sqrt{\frac{6}{\pi e^3}}\right\}.$$

Hence (36.11)' describes (36.11).

Now we sestimate $r(\lambda)$ by virtue of (36.11)'. We obtain, for $0 < \lambda \leqq \lambda_0$, an estimate of $r(\lambda)$:

*Mikusiński [5], p. 225.

$$|r(\lambda)| \leqq |E|*\left[3\sqrt{\frac{6}{\pi e^3}}\right]\left\{\sum_{\nu=1}^{\infty}\frac{1/\alpha^2}{(2\nu\lambda_0+\lambda)^2}+\frac{1/\alpha^2}{(2\nu\lambda_0-\lambda)^2}\right\}$$

$$\leqq \left[3\sqrt{\frac{6}{\pi e^3}}\frac{2}{\alpha^2\lambda_0^2}\sum_{\nu=1}^{\infty}\frac{1}{(2\nu-1)^2}\right]|E|*\{1\}$$

$$\leqq \left[\frac{K}{\alpha^2\lambda_0^2}\right]\left\{\int_0^t |E|(u)\,du\right\} \qquad (K \text{ is a positive constant}),$$

where $|E|$ means the absolute value of the function $E$, and $*$ denotes convolution. More precisely, we have obtained

$$\begin{cases} |r(\lambda,t)| \leqq \dfrac{K}{\alpha^2\lambda_0^2}\displaystyle\int_0^t |E(u)|\,du \qquad (0 < \lambda \leqq \lambda_0,\ 0 \leqq t), \\ K = 6\sqrt{\dfrac{6}{\pi e^3}}\displaystyle\sum_{\nu=1}^{\infty}\frac{1}{(2\nu-1)^2}. \end{cases} \tag{36.13}$$

Therefore, we have

PROPOSITION 35. In (36.13), the error $|r(\lambda,t)|$ tends to $0$ as $\lambda_0$ tends to $\infty$. Thus, for the *infinitely long Thomson cable* with $\alpha = \sqrt{RC}$ and $E = \{E_0\}$, we may take

$$U(\lambda) = \frac{E_0}{s}e^{-\sqrt{RC}\lambda\sqrt{s}}. \tag{36.14}$$

That is, by virtue of (27.14),

$$U(\lambda,t) = E_0\left\{\operatorname{Cerf}\frac{\lambda}{2}\sqrt{\frac{RC}{t}}\right\}^*. \tag{36.15}$$

Thus, by $L = G = 0$, (33.1), and (36.14), we obtain

$$I(\lambda) = -\frac{I}{R}U'(\lambda) = E_0\sqrt{C/R}\,\frac{I}{\sqrt{s}}e^{-\lambda\sqrt{RC}\sqrt{s}}; \tag{36.16}$$

that is, by (27.13),

$$I(\lambda,t) = E_0\left\{\sqrt{\frac{C}{\pi Rt}}\exp\left(-\frac{RC\lambda^2}{4t}\right)\right\}. \tag{36.17}$$

EXERCISES FOR §36. Prove the following inequalities.

(α) $\dfrac{1}{s^2+a^2} \leqq \dfrac{1}{as}$ $(a > 0)$

*Since $\operatorname{Cerf} t = (2/\sqrt{\pi})\int_t^{\infty} e^{-t^2}dt$, the initial condition $\lim_{t\downarrow 0} U(\lambda,t) = 0$ (when $\lambda > 0$) holds.

(ß) $\frac{1}{\sqrt{s}} e^{-\lambda\sqrt{s}} \leqq \frac{1}{\sqrt{\pi e}} \frac{\sqrt{2}}{\lambda s} \quad (\lambda > 0).$

## §37. CONCRETE REPRESENTATIONS OF $\exp(-\lambda\sqrt{\alpha s+\beta})$, $\exp(-\lambda\sqrt{\alpha s^2+\beta s})$ AND $\exp(-\lambda((s+\alpha)^2-\beta^2)^{1/2})$

For the telegraph equation

$$U''(\lambda) = (Ls + R)(Cs + G)U(\lambda),$$

we have the following cases, excepting those of §34, §35 and §36.

Case (i) $L = 0 \quad (R > 0, \quad C > 0, \quad G > 0)$:

$$\exp(-\lambda(RCs + RG)^{1/2}).$$

Case (ii) $G = 0 \quad (L > 0, \quad R > 0, \quad C > 0)$:

$$\exp(-\lambda(LCs^2 + RCs)^{1/2}).$$

Case (iii) $L > 0, \quad R > 0, \quad C > 0, \quad G > 0$:

$$\exp(-\lambda(LC[(s+\alpha)^2 - \beta^2])^{1/2}), \quad \text{where}$$
$$\alpha = \frac{1}{2}\left(\frac{R}{L} + \frac{G}{C}\right), \quad \beta = \frac{1}{2}\left(\frac{R}{L} - \frac{G}{C}\right)$$

since

$$(Ls+R)(Cs+G) = LC\left(\frac{1}{2}\left(\frac{R}{L} + \frac{G}{C}\right) + s + \frac{1}{2}\left(\frac{R}{L} - \frac{G}{C}\right)\right)$$
$$\times \left(\frac{1}{2}\left(\frac{R}{L} + \frac{G}{C}\right) + s - \frac{1}{2}\left(\frac{R}{L} - \frac{G}{C}\right)\right)$$
$$= \frac{1}{LC}(CR+LCs)(LG+LCs) = (R+Ls)(G+Cs).$$

Therefore, the corresponding exponential hyperfunctions come in as solutions of the telegraph equation.

In this section, we shall make use of the Mikusiński mapping $T^\alpha$ of $C/C$ into $C/C$ (introduced in §20) to obtain concrete representation of the above exponential hyperfunctions as functions of $t$.

We already know (see §20) the following:

$$T^\alpha a = T^\alpha\{a(t)\} = \{e^{\alpha t}a(t)\} \qquad (a \in C), \tag{37.1}$$

$$\begin{cases} T^\alpha(a+b) = T^\alpha a + T^\alpha b \\ T^\alpha(ab) = (T^\alpha a)\cdot(T^\alpha b) \end{cases} \qquad (a,b \in C), \tag{37.2}$$

$$T^{\alpha} \frac{q}{p} = \frac{T^{\alpha}q}{T^{\alpha}p} \qquad (p,q \in C/C), \tag{37.3}$$

$$T^{\alpha}(ab) = (T^{\alpha}a)\cdot(T^{\alpha}b) \qquad (a,b \in C/C), \tag{37.4}$$

$$\begin{cases} T^{\alpha}[\gamma] = [\gamma]; \text{ in particular, } T^{\alpha}I = I, \text{ and} \\ T^{\alpha}([\gamma]a) = [\gamma]T^{\alpha}a \qquad (a \in C/C). \end{cases} \tag{37.5}$$

We have further that

$$T^{\alpha}s = T^{\alpha} \frac{I}{\{1\}} = \frac{I}{\{e^{\alpha t}\}} = s - [\alpha] = s - \alpha. \tag{37.6}$$

and, more generally,

$$T^{\alpha}s^n = (s-\alpha)^n \qquad (n = \cdots, -3, -2, -1, 0, 1, 2, 3, \cdots). \tag{37.7}$$

PROOF: If $n > 0$, we have

$$T^{\alpha}s^n = (s - \alpha)^n$$

by (37.6) and (37.4). If $n < 0$, we have

$$T^{\alpha}s^n = \frac{T^{\alpha}I}{T^{\alpha}s^{-n}} = \frac{I}{(s-\alpha)^{-n}} = (s-\alpha)^n.$$

If $n = 0$, we have

$$T^{\alpha}s^o = T^{\alpha}I = I = s^o.$$

If $R(s)$ is a rational function of $s$, then

$$T^{\alpha}R(s) = R(s - \alpha). \tag{37.8}$$

PROOF: We have

$$R(s) = \frac{Q(s)}{P(s)}, \quad \text{where } Q(s) \text{ and } P(s) \text{ are polynomials in } s.$$

Thus

$$T^{\alpha}R(s) = \frac{T^{\alpha}Q(s)}{T^{\alpha}P(s)} = \frac{Q(s-\alpha)}{P(s-\alpha)} = R(s-\alpha),$$

since

$$T^{\alpha}Q(s) = Q(s - \alpha),$$

as may be seen from (37.4), (37.5) and (37.7).

(37.7) is generalized as follows.

$$T^{\alpha}\frac{I}{s^{\lambda}} = \frac{I}{(s-\alpha)^{\lambda}} \qquad (-\infty < \lambda < \infty). \tag{37.9}$$

PROOF: Let $n$ be a natural number such that $\lambda+n > 1$. Then

$$\frac{I}{s^{\lambda}} = h^{\lambda} = \frac{h^{\lambda+n}}{h^{n}} = \frac{\Gamma(\lambda+n)^{-1}t^{\lambda+n-1}}{\Gamma(n)^{-1}t^{n-1}}$$

and so

$$T^{\alpha}\frac{I}{s^{\lambda}} = T^{\alpha}h^{\lambda} = \frac{T^{\alpha}h^{\lambda+n}}{T^{\alpha}h^{n}} = \frac{\Gamma(\lambda+n)^{-1}t^{\lambda+n-1}e^{\alpha t}}{\Gamma(n)^{-1}t^{n-1}e^{\alpha t}}$$

$$= \frac{I}{(s-\alpha)^{\lambda+n}}\left(\frac{I}{(s-\alpha)^{n}}\right)^{-1} = \frac{I}{(s-\alpha)^{\lambda}}$$

by (10.20).

EXAMPLE 37.1.

$$T^{\alpha}\sqrt{s} = \sqrt{s-\alpha}, \qquad T^{\alpha}\frac{I}{\sqrt{s}} = \frac{I}{\sqrt{s-\alpha}} \tag{37.10}$$

and

$$\begin{aligned} T^{\alpha}\sqrt{s^2-\beta^2} &= (T^{\alpha}\sqrt{s-\beta})(T^{\alpha}\sqrt{s+\beta}) \\ &= \sqrt{s-\beta-\alpha}\,\sqrt{s+\beta-\alpha} = \sqrt{(s-\alpha)^2-\beta^2}\,. \end{aligned} \tag{37.11}$$

If $f(\lambda)$ has the generalized derivative $f'(\lambda)$, then

$$(T^{\alpha}f(\lambda))' = T^{\alpha}f'(\lambda). \tag{37.12}$$

PROOF: There exists an element $a \in C/C$ with $a \neq 0$ such that $af(\lambda) = \{f_1(\lambda,t)\}$ is continuously differentiable with respect to $\lambda$ and

$f'(\lambda) = a^{-1}\{\frac{\partial}{\partial\lambda}f_1(\lambda,t)\}$. Hence, by (37.4),

$$T^{\alpha}f(\lambda) = T^{\alpha}a^{-1}\{f_1(\lambda,t)\} = (T^{\alpha}a^{-1})(T^{\alpha}\{f_1(\lambda,t)\}).$$

Thus

$$\begin{aligned} (T^{\alpha}f(\lambda))' &= (T^{\alpha}a^{-1})(T^{\alpha}\{\tfrac{\partial}{\partial\lambda}f_1(\lambda,t)\}) \\ &= T^{\alpha}(a^{-1}\{\tfrac{\partial}{\partial\lambda}f_1(\lambda,t)\}) \\ &= T^{\alpha}f'(\lambda). \end{aligned}$$

As a Corollary, we have

$$T^{\alpha}\exp(\lambda w) = \exp(\lambda T^{\alpha}w). \tag{37.13}$$

PROOF: We have

$$(T^{\alpha}\exp(\lambda w))' = T^{\alpha}(\exp(\lambda w))' = T^{\alpha}(w\exp(\lambda w))$$
$$= T^{\alpha}w\cdot T^{\alpha}\exp(\lambda w),$$
$$T^{\alpha}\exp(0\cdot w) = T^{\alpha}I = I.$$

Hence, by Proposition 24, we must have (37.13).

APPLICATIONS OF THE MAPPING $T^{\alpha}$.

$$e^{-\lambda\sqrt{s-\alpha}} = T^{\alpha}e^{-\lambda\sqrt{s}} = \left\{\frac{\lambda}{2\sqrt{\pi t^3}}\exp\left(\alpha t - \frac{\lambda^2}{4t}\right)\right\} \tag{37.14}$$

by (27.5).

$$\frac{I}{\sqrt{s-\alpha}}e^{-\lambda\sqrt{s-\alpha}} = T^{\alpha}\frac{I}{\sqrt{s}}e^{-\lambda\sqrt{s}} = \left\{\frac{1}{\sqrt{\pi t}}\exp\left(\alpha t - \frac{\lambda^2}{4t}\right)\right\} \tag{37.15}$$

by (27.3).

For any real number $\alpha$ and any positive number $\beta$, we have

$$\left\{\mathrm{Erf}\left(\alpha\sqrt{t} + \frac{\beta}{\sqrt{t}}\right)\right\} = \frac{1}{s}e^{-2\alpha\beta}\left(\frac{\alpha}{\sqrt{s+\alpha^2}} - 1\right)e^{-2\beta\sqrt{s+\alpha^2}} + \frac{I}{s}, \tag{37.16}$$

and

$$\left\{\mathrm{Erf}\left(\alpha\sqrt{t} - \frac{\beta}{\sqrt{t}}\right)\right\} = \frac{1}{s}e^{2\alpha\beta}\left(\frac{\alpha}{\sqrt{s+\alpha^2}} + 1\right)e^{-2\beta\sqrt{s+\alpha^2}} - \frac{I}{s}. \tag{37.17}$$

<u>Proof</u>: For (37.16): Writing

$$f(t) = \frac{2}{\sqrt{\pi}}\int_0^{\alpha\sqrt{t}+\beta/\sqrt{t}} e^{-u^2}du = \mathrm{Erf}\left(\alpha\sqrt{t} + \frac{\beta}{\sqrt{t}}\right),$$

we have

$$f'(t) = e^{-2\alpha\beta}e^{-\alpha^2 t}\left(\frac{\alpha}{\sqrt{\pi t}}e^{-\beta^2/t} - \frac{\beta}{\sqrt{\pi t^3}}e^{-\beta^2/t}\right).$$

Hence, by (27.3) and (27.5),

$$\{f'(t)\} = e^{-2\alpha\beta}T_{-\alpha^2}\left(\frac{\alpha}{\sqrt{s}}\, e^{-2\beta\sqrt{s}} - e^{-2\beta\sqrt{s}}\right)$$

$$= e^{-2\alpha\beta}\left(\frac{\alpha}{\sqrt{s+\alpha^2}} - 1\right)e^{-2\beta\sqrt{s+\alpha^2}} .$$

Furthermore,

$$f' = sf - s[1]$$

by

$$f(+0) = \lim_{t\downarrow 0} f(t) = 1;$$

that is,

$$f = \frac{f'}{s} + \frac{I}{s},$$

proving that (37.16) is valid.

Similarly, writing

$$g(t) = \frac{2}{\sqrt{\pi}}\int_0^{\alpha\sqrt{t}-\beta/\sqrt{t}} e^{-u^2}\,du = \mathrm{Erf}\left(\alpha\sqrt{t} - \frac{\beta}{\sqrt{t}}\right),$$

we have

$$\{g'(t)\} = e^{2\alpha\beta}\left(\frac{\alpha}{\sqrt{s+\alpha^2}} + 1\right)e^{-2\beta\sqrt{s+\alpha^2}},$$

whence follows (37.17) by virtue of $g(+0) = -1$.

From (37.16) and (37.17), we easily obtain

$$\frac{I}{s}\, e^{-2\beta\sqrt{s+\alpha^2}} = \left\{\frac{1}{2}\, e^{-2\alpha\beta}\mathrm{Erf}\left(\alpha\sqrt{t} - \frac{\beta}{\sqrt{t}}\right) - \frac{1}{2}\, e^{2\alpha\beta}\mathrm{Erf}\left(\alpha\sqrt{t} + \frac{\beta}{\sqrt{t}}\right) - \frac{e^{2\alpha\beta}+e^{-2\alpha\beta}}{2}\right\} \tag{37.18}$$

and

$$\frac{\alpha}{s\sqrt{s+\alpha^2}}\, e^{-2\beta\sqrt{s+\alpha^2}} = \left\{\frac{1}{2}\, e^{-2\alpha\beta}\mathrm{Erf}\left(\alpha\sqrt{t} - \frac{\beta}{\sqrt{t}}\right) + \frac{1}{2}\, e^{2\alpha\beta}\mathrm{Erf}\left(\alpha\sqrt{t} + \frac{\beta}{\sqrt{t}}\right) - \frac{e^{2\alpha\beta}-e^{-2\alpha\beta}}{2}\right\}. \tag{37.19}$$

Next we have

$$s - \sqrt{(s-\alpha)^2 + \beta^2} = \alpha + T^{\alpha}(s - \sqrt{s^2 + \beta^2})$$

by (37.6) and (37.11). Hence, by (28.15) and (28.17), we obtain

$$\exp(\lambda(s - \sqrt{(s-\alpha)^2+\beta^2}) = e^{\alpha\lambda} - \left\{\frac{\lambda}{\sqrt{t^2+2\lambda t}} e^{(\lambda+t)}\beta J_1(\beta\sqrt{t^2+2\beta t})\right\} \quad (37.20)$$

and

$$\frac{\exp(\lambda(s-\sqrt{(s-\alpha)^2+\beta^2}))}{\sqrt{(s-\alpha)^2 + \beta^2}} = \left\{e^{\alpha(\lambda+t)} J_0(\beta\sqrt{t^2+2\lambda t})\right\}. \quad (37.21)$$

Replacing, in formulas (37.20)-(37.21), $\alpha$ by $-\alpha$ and $\beta$ by $i\alpha$, we obtain

$$\exp(\lambda(s-\sqrt{s^2+2\alpha s}) = e^{-\alpha\lambda} - \left\{\frac{\lambda}{\sqrt{t^2+2\lambda t}} e^{-\alpha(\lambda+t)} i\alpha J_1(i\alpha\sqrt{t^2+2\lambda t})\right\} \quad (37.22)$$

and

$$\frac{\exp(\lambda(s-\sqrt{s^2+2\alpha s}))}{\sqrt{s^2+2\alpha s}} = \left\{e^{-\alpha(\lambda+t)} J_0(i\alpha\sqrt{t^2+2\lambda t})\right\}. \quad (37.23)$$

Multiplying (37.20), (37.21), (37.22), and (37.23) by the shift operator $e^{-\lambda s}$ replaces $t$ by $(t-\lambda)$ and so we have

$$\exp(-\lambda\sqrt{(s-\alpha)^2+\beta^2}) = e^{\alpha\lambda}e^{-\lambda s} - \left\{\begin{array}{ll} 0, & 0 \leqq t < \lambda, \\ \dfrac{\lambda}{\sqrt{t^2-\lambda^2}} e^{\alpha t}\beta J_1(\beta\sqrt{t^2-\lambda^2}), & 0 \leqq \lambda < t \end{array}\right\}, \quad (37.24)$$

$$\frac{\exp(-\lambda\sqrt{(s-\alpha)^2+\beta^2})}{\sqrt{(s-\alpha)^2+\beta^2}} = \left\{\begin{array}{ll} 0, & 0 \leqq t < \lambda \\ e^{\alpha t} J_0(\beta\sqrt{t^2-\lambda^2}), & 0 \leqq \lambda < t \end{array}\right\}, \quad (37.25)$$

$$\exp(-\lambda\sqrt{s^2+2\alpha s}) = e^{-\alpha\lambda}e^{-\lambda s} - \left\{\begin{array}{ll} 0, & 0 \leqq t < \lambda, \\ \dfrac{\lambda}{\sqrt{t^2-\lambda^2}} e^{-\alpha t} i\alpha J_1(i\alpha\sqrt{t^2-\lambda^2}), & 0 \leqq \lambda < t \end{array}\right\}, \quad (37.26)$$

and

$$\frac{\exp(-\lambda\sqrt{s^2+2\alpha s})}{\sqrt{s^2+2\alpha s}} = \left\{\begin{array}{ll} 0, & 0 \leqq t < \lambda \\ e^{-\alpha t}J_0(i\alpha\sqrt{t^2-\lambda^2}), & 0 \leqq \lambda < t \end{array}\right\}. \tag{37.27}$$

Therefore, we have obtained concrete representations of the cases (i) and (ii) mentioned at the beginning of this section, by (37.14) and (37.26) respectively. As for the case (iii), we obtain the following:

$$\exp(-\lambda\sqrt{(s-\alpha)^2-\beta^2}) = e^{\alpha\lambda}e^{-\lambda s}$$

$$- \left\{\begin{array}{ll} 0, & 0 \leqq t < \lambda, \\ \dfrac{\lambda}{\sqrt{t^2-\lambda^2}} e^{\alpha t} i\beta J_1(i\beta\sqrt{t^2-\lambda^2}), & 0 \leqq \lambda < t \end{array}\right\}, \tag{37.28}$$

which can be derived by replacing $\beta$ in (37.24) by $i\beta$.

## §38. A CABLE WITHOUT SELF-INDUCTION

This is the case

$$L = 0 \quad (R > 0,\ C > 0,\ G > 0).$$

The hyperfunction equation of the cable is

$$U''(\lambda) = (RCs + RG)U(\lambda) \qquad (0 \leqq \lambda \leqq \lambda_0) \tag{38.1}$$

with the $\lambda$-boundary condition

$$U(0) = E = \{E_0\}, \qquad U(\lambda_0) = 0. \tag{38.2}$$

Thus, putting

$$\alpha = \sqrt{RC}, \quad \gamma = RG/RC = G/C, \tag{38.3}$$

the solution will be given by

$$\left\{\begin{array}{l} U(\lambda) = a\exp(-\alpha\lambda\sqrt{s+\gamma}) + b\exp(\alpha\lambda\sqrt{s+\gamma}), \\ a \in C/C, \quad b \in C/C. \end{array}\right.$$

Hence, by (38.2), we have

$$a+b = E, \quad a\exp(-\alpha\lambda_0\sqrt{s+\gamma}) + b\exp(\alpha\lambda_0\sqrt{s+\gamma}) = 0,$$

and so

$$a = \frac{E}{I-e^{-2\alpha\lambda_0\sqrt{s+\gamma}}}, \quad b = -\frac{e^{-2\alpha\lambda_0\sqrt{s+\gamma}E}}{I-e^{-2\alpha\lambda_0\sqrt{s+\gamma}}},$$

$$U(\lambda) = \frac{(e^{-\alpha\lambda\sqrt{s+\gamma}} - e^{-\alpha(2\lambda_0-\lambda)\sqrt{s+\gamma}})}{I-e^{-2\alpha\lambda_0\sqrt{s+\gamma}}} E.$$

In

$$T^{\gamma}U(\lambda) = \frac{e^{-\alpha\lambda\sqrt{s}} - e^{-\alpha(2\lambda_0-\lambda)\sqrt{s}}}{I-e^{-2\alpha\lambda_0\sqrt{s}}}\hat{E} \qquad (\hat{E} = T^{\gamma}E),$$

we let $\lambda_0 \to \infty$. Then as in the case of the infinitely long cable (§36), we may take

$$T^{\gamma}U(\lambda) = e^{-\alpha\lambda\sqrt{s}}\,\hat{E}.$$

Thus, since $T^{-\gamma}T^{\gamma} = I$,

$$U(\lambda) = T^{-\gamma}(e^{-\alpha\lambda\sqrt{s}}\,\hat{E}) = e^{-\alpha\lambda\sqrt{s+\gamma}}\,E;$$

that is, when $E = \{E_0\}$,

$$U(\lambda) = \frac{[E_0]}{s}\exp\left(-RC\lambda(\sqrt{s + G/C})\right). \tag{38.5}$$

Therefore, by (37.18), we have, when $E = \{E_0\}$,

$$U(\lambda,t) = \frac{E_0}{2}\left[e^{-\lambda\sqrt{RG}}\,\mathrm{Erf}\left(\sqrt{Gt/C} - \frac{\lambda}{2}(\sqrt{RC/t})\right)\right.$$
$$\left. - e^{\lambda\sqrt{RG}}\,\mathrm{Erf}\left(\sqrt{Gt/C} + \frac{\lambda}{2}(\sqrt{RC/t})\right) - (e^{\lambda\sqrt{RG}} + e^{-\lambda\sqrt{RG}})\right]. \tag{38.6}$$

Next, for $I(\lambda)$, we obtain, by (33.3),

$$I(\lambda) = \frac{-1}{R}U'(\lambda) = \frac{E_0\sqrt{RCs + RG}}{Rs}\exp(-\lambda\sqrt{RCs + RG})$$
$$= E_0\left(\frac{C}{\sqrt{RCs+RG}} + \frac{G}{s\sqrt{RCs+RG}}\right)\exp(-\lambda\sqrt{RCs + RG}), \tag{38.7}$$

whence, by (37.14), (37.15) and (37.19),

$$I(\lambda,t) = \sqrt{C/R}\,\frac{E_0}{\sqrt{\pi t}}\exp\left(-\frac{G}{C}t - \frac{RC}{4t}\lambda^2\right)$$

$$+ \sqrt{G/R}\,\frac{E_0}{2}\left[e^{-\lambda\sqrt{RG}}\,\mathrm{Erf}\left(\sqrt{Gt/C} - \frac{\lambda}{2}(\sqrt{RC/t})\right)\right. \qquad (38.8)$$

$$\left. + e^{\lambda\sqrt{GR}}\,\mathrm{Erf}\left(\sqrt{Gt/C} + \frac{\lambda}{2}\sqrt{RC/t}\right) - (e^{\lambda\sqrt{RG}} - e^{-\lambda\sqrt{RG}})\right].$$

## §39. A CABLE WITHOUT-LEAK CONDUCTANCE

This is the case

$G = 0 \quad (L > 0,\ C > 0,\ R > 0).$

In many practical calculations, the leakage of the current is so small that it can be neglected. Thus the equation is

$$U''(\lambda) = (LCs^2 + RCs)U(\lambda), \quad U(0) = E, \qquad (39.1)$$

under, of course, the assumption that at the initial instant $t = 0$ there is no voltage and no current on the cable.

The general solution of the hyperfunction equation (39.1) is

$$U(\lambda) = a\exp(-\lambda\sqrt{LCs^2 + RCs}) + b\exp(\lambda\sqrt{LCs^2 + RCs}),$$
$$a \in C/C, \quad b \in C/C. \qquad (39.2)$$

We shall be concerned with such a *hyperfunction solution* $U(\lambda)$ *which can be represented as a numerical-valued function* $U(\lambda,t)$ of $\lambda > 0$ and $t \geqq 0$. This situation is essential, since otherwise the $U(\lambda)$ *cannot be calculated numerically.*

Assuming that *the cable is infinitely long,* we have

PROPOSITION 36. If the solution (39.2) can be represented as a numerical-valued function $U(\lambda,t)$ in the domain

$0 < \lambda < \infty, \quad 0 \leqq t < \infty,$

then the coefficient $b$ must be $0$.

PROOF:* By the binomial expansion (14.3), we have

---

*Mikusiński [5], p. 258.

$$
\begin{aligned}
(LCs^2 + RCs)^{1/2} &= \sqrt{LC}\, s\left(I + \frac{R}{Ls}\right)^{1/2} \\
&= \sqrt{LC}\, s\left(I + \frac{R}{2Ls} + \sum_{n=2}^{\infty} \binom{1/2}{n} \left(\frac{R}{L}\right)^n h^n\right) \qquad (39.3) \\
&= \sqrt{LC}\, s + \frac{\sqrt{LC}\, R}{2L} + f = \sqrt{LC}\, s + \alpha + f,
\end{aligned}
$$

where

$$\alpha = \frac{\sqrt{C}\, R}{2\sqrt{L}}, \quad f = \{f(t)\} = \left\{\sum_{n=2}^{\infty} \binom{1/2}{n} \left(\frac{R}{L}\right)^n h^{n-1}\right\} \in C.$$

By (39.2), (39.3) and (28.10), we have

$$U(\lambda) - a \exp(-\lambda\sqrt{LC}\, s) e^{-\lambda\alpha} e^{-\lambda f} = b \exp(\lambda\sqrt{LC}\, s) e^{\lambda\alpha} e^{\lambda f}. \qquad (30.4)$$

Putting

$$a = \frac{q}{p}, \quad b = \frac{q_1}{p_1} \quad (p, q, p_1, q_1 \in C),$$

and then nultiplying (39.4) by $pp_1 e^{-\lambda\alpha} e^{-\lambda f}$, we obtain

$$
\begin{aligned}
pp_1 e^{-\alpha\lambda} e^{-\lambda f} U(\lambda) &- qp_1 e^{-2\lambda\alpha} e^{-2\lambda f} \exp(-\lambda\sqrt{LC}\, s) \\
&= pq_1 \exp(\lambda\sqrt{LC}\, s).
\end{aligned} \qquad (39.5)
$$

Since $f$ is a numerical-valued continuous function of $t$, $pp_1 e^{-\alpha\lambda} e^{-\lambda f} U(\lambda)$ is also a numerical-valued continuous function of $\lambda > 0$ and $t \geqq 0$. Furthermore, since $\exp(-\lambda\sqrt{LC}\, s)$ is a shift operator,

$$\exp(-\lambda\sqrt{LC}\, s)(qp_1 e^{-2\lambda\alpha} e^{-2\lambda f})$$

is a numerical-valued function in $K[0,\infty)$ when $\lambda > 0$. Thus, by (39.5),

$$\left\{\begin{array}{l} \exp(\lambda\sqrt{LC}\, s) pq_1 \quad \text{is a numerical-valued function} \\ \text{of } \lambda \text{ and } t \text{ in the domain } 0 < \lambda < \infty, \ 0 \leqq t. \end{array}\right. \qquad (39.6)$$

On the other hand, for any $\lambda_0 > 0$, and $\lambda > 0$ with $\lambda_0 > \lambda > 0$, we have, denoting

$$\exp(\lambda_0 \sqrt{LC}\, s) pq_1 = \{g(t)\},$$

that

$$\begin{aligned}\exp(\lambda\sqrt{LC}\ s)pq_1 &= \exp(-(\lambda_0-\lambda)\sqrt{LC}\ s)\exp(\lambda_0\sqrt{LC}\ s)pq_1 \\ &= \exp(-(\lambda_0-\lambda)\sqrt{LC}\ s)\{g(t)\} \\ &= 0 \quad \text{when} \quad 0 \leqq t < (\lambda_0-\lambda)\sqrt{LC},\end{aligned} \tag{39.7}$$

since $\exp(-(\lambda_0-\lambda)\sqrt{LC}\ s)$ with $(\lambda_0-\lambda) > 0$ is a shift operator.

Hence, by the assumption that the cable is infinitely long, we may put

$$\exp(\lambda\sqrt{LC}\ s)pq_1 = 0$$

by letting $\lambda_0 \to \infty$. Thus

$$b\ \exp(\lambda\sqrt{CL}\ s) = \frac{I}{pp_1}\exp(\lambda\sqrt{CL}\ s)pq_1 = 0 \quad \text{for} \quad \lambda > 0.$$

But, as an exponential hyperfunction, $\exp(\lambda\sqrt{LC}\ s) \in \mathcal{C}/\mathcal{C}$ does not vanish.* This proves that

$$b = 0.$$

Therefore, combined with (39.2) and the $\lambda$-initial condition $U(0) = E$, we obtain

$$U(\lambda) = E\ \exp(-\lambda\sqrt{LCs^2 + RCs}). \tag{39.8}$$

If $E$ is of the form $E = \{E_0\}$, then

$$U(\lambda) = \frac{E_0}{s}\exp(-\lambda\sqrt{LCs^2 + RCs}), \tag{39.9}$$

and, by (33.3),

$$\begin{aligned}I(\lambda) &= -\frac{U'(\lambda)}{Ls+R} = \frac{E_0/s}{Ls+R}\sqrt{LCs^2 + RCs}\ \exp(-\lambda\sqrt{LCs^2 + RCs}) \\ &= \frac{E_0C}{\sqrt{LCs^2 + RCs}}\exp(-\lambda\sqrt{LCs^2 + RCs}).\end{aligned} \tag{39.10}$$

Hence, by putting

$$\alpha' = \frac{1}{2}\frac{R}{L},$$

we see that

---

*See Proposition 24 (§28).

$$U(\lambda) = \frac{E_0}{s}\exp(-\lambda\sqrt{LC}\sqrt{s^2+2\alpha' s})$$

$$= E_0 h \exp(-\lambda\sqrt{LC}\sqrt{s^2+2\alpha' s}).$$

Therefore, by (37.26),

$$U(\lambda,t) = E_0\Big(e^{-\alpha'\sqrt{LC}\lambda}(t - \sqrt{LC}\,\lambda)$$

$$-\frac{1}{2}\int_{\sqrt{LC}\,\lambda}^{t} \frac{\sqrt{LC}\,\lambda\cdot R/L}{\sqrt{\tau^2-LC\lambda^2}} e^{-\alpha'\tau} iJ_1(i\alpha'\sqrt{\tau^2-LC\lambda^2})d\tau\Big), \qquad (39.11)^*$$

when $0 \leqq \sqrt{LC}\,\lambda < t$.

Similarly, by (37.27) and (39.10),

$$I(\lambda,t) = E_0\sqrt{C/L}\, e^{-\alpha' t} J_0(i\alpha'\sqrt{t^2-LC\lambda^2})$$

when $0 \leqq \sqrt{LC}\,\lambda < t$. Moreover, both $U(\lambda,t)$ and $I(\lambda,t)$ vanish when $0 \leqq t < \sqrt{LC}\,\lambda$.

## §40. THE CASE WHERE ALL THE FOUR PARAMETERS ARE POSITIVE

This is the case

$$L > 0, \quad R > 0, \quad C > 0, \quad G > 0.$$

Introducing, as in §37, the notation

$$\alpha = \frac{1}{2}\left(\frac{R}{L} + \frac{G}{C}\right), \quad \beta = \frac{1}{2}\left(\frac{R}{L} - \frac{G}{C}\right),$$

the hyperfunction equation of the cable is

$$U''(\lambda) = LC[(s+\alpha)^2 - \beta^2]U(\lambda); \quad U(0) = E = \{E_0\}. \qquad (40.1)$$

As in the preceding section §39, we assume that *the cable is infinitely long*. Then the solution $U(\lambda)$ of (40.1), which can be represented as a numerical-valued function $U(\lambda,t)$ in the domain $0 < \lambda < \lambda_0$, $0 \leqq t < \infty$, is given by

$$U(\lambda) = \frac{E_0}{s}\exp(-\lambda\sqrt{LC}((s+\alpha)^2 - \beta^2)^{1/2}). \qquad (40.2)$$

By (33.3), we have

*Compare with that on pg. 250 of J. Mikusiński, [5].

$$I(\lambda) = \frac{-U'(\lambda)}{Ls + R}$$

$$= E_0 \frac{\sqrt{LC}((s+\alpha)^2 - \beta^2)^{1/2}}{s(Ls + R)} \exp(-\lambda\sqrt{LC}((s+\alpha^2) - \beta^2)^{1/2}),$$

so that

$$I(\lambda) = E_0 \sqrt{C/L}(I + \frac{G}{Cs}) \frac{\exp(-\lambda\sqrt{LC}((s+\alpha)^2 - \beta^2)^{1/2})}{((s+\alpha)^2 - \beta^2)^{1/2}} \tag{40.3}$$

by making use of

$$\sqrt{C/L}(I + \frac{G}{Cs}) \times \frac{s(Ls+R)}{\sqrt{LC}} = \frac{(Cs + G)(Ls + R)}{LC} = (s+\alpha)^2 - \beta^2.$$

Therefore, by (37.24) and (40.2),

$$U(\lambda,t) = E_0\Big(e^{-\alpha\lambda\sqrt{LC}}(t - \sqrt{LC}\,\lambda) - \int_{\sqrt{LC}\,\lambda}^{t} \frac{\lambda\sqrt{LC}}{\sqrt{\tau^2-\lambda^2 LC}} e^{-\alpha\tau} i\beta J_1(i\beta\sqrt{\tau^2-LC\lambda^2})\,d\tau\Big), \tag{40.4*}$$

when $0 \leqq \sqrt{LC}\,\lambda < t$.

Also, by (27.25) and (40.3),

$$I(\lambda,t) = E_0 \sqrt{C/L}\Big((e^{-\alpha t}J_0(i\beta\sqrt{t^2-LC\lambda^2}) + \frac{G}{C}\int_{\sqrt{LC}\,\lambda}^{t} e^{-\alpha\tau}J_0(i\beta\sqrt{\tau^2-LC\lambda^2})\,d\tau\Big) \tag{40.5}$$

when $0 \leqq \sqrt{LC}\,\lambda < t$. Moreover, both $U(\lambda,t)$ and $I(\lambda,t)$ vanish when $0 \leqq t < \sqrt{LC}\,\lambda$.

REMARK 40.1. Formulas (40.4) and (40.5) tell us that, with the growth of $\lambda$, the greater the number $\alpha$ the faster the voltage and the current decrease. Thus $\alpha$ is called a *damping coefficient*. The number $\beta$ is called as the *deformation coefficient*, since if $\beta = 0$ then as $\lambda$ increases the wave is only damped and not deformed at all.

---

*Compare with that on pg. 260 of J. Mikusiński [5].

# Chapter XI
# Heat Equation

## §41. THE TEMPERATURE OF A HEAT-CONDUCTING BAR

Let us imagine that a bar of length $\lambda_0$ is placed along the $\lambda$-axis, the abcissa of the left end of the bar being $\lambda = 0$ and the right end $\lambda = \lambda_0$. Let $k$ denote the *heat conductivity*, $c$ the *specific heat*, and $\rho$ the *mass density* of the bar. Furthermore, let the lateral surface of the bar be perfectly insulated so that heat can flow in and flow out only through the ends of the bar. If we denote by $z(\lambda,t)$ the temperature at the point of the bar at abcissa $\lambda$ at the instant $t$, then the heat equation in the bar is

$$z_{\lambda\lambda}(\lambda,t) = \alpha^2 z_t(\lambda,t) \quad (\alpha = \sqrt{c\rho/k}). \tag{41.1}$$

Assume that at the initial instant $t = 0$ the temperature throughout the bar is 0, i.e.,

$$z(\lambda,0) = 0 \quad (0 \leqq \lambda < \lambda_0), \tag{41.2}$$

and that the temperatures at the ends of the bar are given by

$$z(0,t) = v_1(t), \quad z(\lambda_0,t) = v_2(t) \qquad (0 \leqq t < \infty). \tag{41.3}$$

By (41.2), we have $z_t = sz$ so that the hyperfunction equation of (41.1) can be written

$$z''(\lambda) = \alpha^2 s z(\lambda). \tag{41.4}$$

In the general solution of (41.4), given by

$$z(\lambda) = ae^{-\alpha\lambda\sqrt{s}} + be^{+\alpha\lambda\sqrt{s}} \quad (a \text{ and } b \in C/C),$$

we shall choose a and b in such a way that (41.3) is satisfied:

$$\begin{cases} z(0) = v_1: \quad a+b = v_1 \\ z(\lambda_0) = v_2: \quad ae^{-\alpha\lambda_0\sqrt{s}} + be^{\alpha\lambda_0\sqrt{s}} = v_2. \end{cases} \tag{41.5}$$

Hence

$$a = \frac{v_1 - v_2 e^{-\alpha\lambda_0\sqrt{s}}}{I - e^{-2\alpha\lambda_0\sqrt{s}}}, \quad b = \frac{-e^{-2\alpha\lambda_0\sqrt{s}} v_1 + e^{-\alpha\lambda_0\sqrt{s}} v_2}{I - e^{-2\alpha\lambda_0\sqrt{s}}}, \tag{41.6}$$

and so

$$\begin{aligned} z(\lambda) &= \frac{e^{-\alpha\lambda\sqrt{s}} - e^{-\alpha(2\lambda_0-\lambda)\sqrt{s}}}{I - e^{-2\alpha\lambda_0\sqrt{s}}} v_1 \\ &+ \frac{e^{-\alpha(\lambda_0-\lambda)\sqrt{s}} - e^{-\alpha(\lambda_0+\lambda)\sqrt{s}}}{I - e^{-2\alpha\lambda_0\sqrt{s}}} v_2 \\ &= z_1(\lambda) + z_2(\lambda). \end{aligned} \tag{41.7}$$

Furthermore, the fact that the hyperfunction solution of (41.3)-(41.4) is uniquely determined by (41.7) shall be proved as in Proposition 30.

Now we have the series expansion

$$\begin{aligned} z_1(\lambda) &= v_1 \cdot \sum_{\nu=0}^{\infty} \left( e^{-\alpha(2\nu\lambda_0+\lambda)\sqrt{s}} - e^{-\alpha(2(\nu+1)\lambda_0-\lambda)\sqrt{s}} \right), \\ z_2(\lambda) &= v_2 \cdot \sum_{\nu=0}^{\infty} \left( e^{-\alpha((2\nu+1)\lambda_0-\lambda)\sqrt{s}} - e^{-\alpha((2\nu+1)\lambda_0+\lambda)\sqrt{s}} \right), \end{aligned} \tag{41.8}$$

since

$$\frac{I}{I - e^{-2\alpha\lambda_0\sqrt{s}}} = I + e^{-2\alpha\lambda_0\sqrt{s}} + \ldots + e^{-2\alpha\lambda_0\nu\sqrt{s}} + \cdots .$$

Hence, by

$$\frac{I}{s} \exp(-\lambda s^{1/2}) = \left\{ \operatorname{Cerf} \frac{\lambda}{2\sqrt{t}} \right\} = \left\{ \frac{2}{\sqrt{\pi}} \int_{\lambda/2\sqrt{t}}^{\infty} e^{-\nu^2} d\nu \right\}, \tag{27.13}$$

we have, when $v_1 = \{E_1\} = \dfrac{[E_1]}{s}$,

$$z_1(\lambda,t) = E_1\left\{\sum_{\nu=0}^{\infty} \frac{2}{\sqrt{\pi}} \int_{\alpha(2\nu\lambda_0+\lambda)/2\sqrt{t}}^{\alpha(2(\nu+1)\lambda_0-\lambda)/2\sqrt{t}} e^{-\nu^2} d\nu\right\} \tag{41.9}$$

$(0 \leqq \lambda < \lambda_0$ and $0 < t < \infty)$.

Similarly we have, when $v_2 = \{E_2\} = \dfrac{[E_2]}{s}$,

$$z_2(\lambda,t) = E_2\left\{\sum_{\nu=0}^{\infty} \frac{2}{\sqrt{\pi}} \int_{\alpha((2\nu+1)\lambda_0-\lambda)/2\sqrt{t}}^{\alpha((2\nu+1)\lambda_0+\lambda)/2\sqrt{t}} e^{-\nu^2} d\nu\right\} \tag{41.10}$$

$(0 \leqq \lambda < \lambda_0$ and $0 < t < \infty)$.

Thus

$$\begin{aligned} z_1(\lambda,0) &= \lim_{t\downarrow 0} z_1(\lambda,t) = 0, \quad z_1(0,t) = E_1, \quad z_1(\lambda_0,t) = 0, \\ z_2(\lambda,0) &= \lim_{t\downarrow 0} z_2(\lambda,t) = 0, \quad z_2(0,t) = 0, \quad z_2(\lambda_0,t) = E_2. \end{aligned} \tag{41.11}$$

Hence the initial condition (at $t = 0$) and the boundary conditions (at $\lambda = 0$ and at $\lambda = \lambda_0$) are satisfied as above, although $z(\lambda,t)$ is not defined as a function of $(\lambda,t)$ at $t = 0$. The same remark applies to $z_2(\lambda,t)$.

## §42. AN INFINITELY LONG BAR

We shall prove the following proposition concerning an *infinitely long bar*.

PROPOSITION 37. Let a hyperfunction solution $z(\lambda)$ of

$$z_1''(\lambda) = \alpha^2 s z_1(\lambda) \quad (\alpha > 0 \quad \text{and} \quad 0 \leqq \lambda < \infty) \tag{42.1}$$

be given by

$$z(\lambda) = ae^{-\alpha\lambda\sqrt{s}} + be^{\alpha\lambda\sqrt{s}} \quad (a,b \in C/C) \tag{42.2}$$

in such a way that it satisfies

$$z(\lambda,0) = 0 \quad (0 \leqq \lambda < \infty), \quad z(0,t) = v_1(t) \in C[0,\infty) \tag{42.3}$$

together with

$$|z(\lambda,t)| \leqq m(t) \quad (0 \leqq \lambda < \infty, \quad 0 \leqq t < \infty). \tag{42.4}$$

Here $z(\lambda,t)$ is a numerical-valued function in the domain

$0 < \lambda < \infty, \quad 0 < t < \infty.$

Then we must have

$$z(\lambda) = v_1 e^{-\alpha\lambda\sqrt{s}} = z_1(\lambda). \tag{42.5}$$

PROOF: From (42.2), we have

$$b = z(\lambda)\, e^{-\alpha\lambda\sqrt{s}} - a e^{-2\alpha\lambda\sqrt{s}}. \tag{42.6}$$

On the other hand, we have

$$0 \leqq e^{-\alpha\lambda\sqrt{s}} \leq \left[3\sqrt{(6/\pi e^3)}\ \frac{1}{\alpha^2\lambda^2}\right]h \qquad (0 < \lambda < \infty \text{ and } 0 < t < \infty) \tag{42.7}$$

by (36.11)'. Hence, by (42.4), we obtain

$$\lim_{\lambda\to\infty} z(\lambda,t)e^{-\alpha\lambda\sqrt{s}} = 0, \qquad \lim_{\lambda\to 0} a e^{-2\alpha\lambda\sqrt{s}} = 0.$$

Thus, by (42.6), we must have $b = 0$. Therefore, by (42.2) and $z(0,t) = v_1(t)$, we obtain (42.5).

COROLLARY. If we further assume that

$$v = \{E_1\} = \frac{[E_1]}{s},$$

we obtain, by (27.14),

$$z(\lambda,t) = E_1\left\{\operatorname{Cerf}\frac{\alpha\lambda}{2\sqrt{t}}\right\} = E_1\left\{\frac{2}{\sqrt{\pi}}\int_{\alpha\lambda/2\sqrt{t}}^{\infty} e^{-\nu^2}\,d\nu\right\}. \tag{42.8}$$

Conversely, it is easy to show that (42.8) satisfies the assumptions of Proposition 37 if we define

$$z(\lambda,0) = \lim_{t\downarrow 0} z(\lambda,t) = 0.$$

AN EXAMPLE OF THE APPLICATION OF (42.8). Let us imagine a long bar of silver with initial temperature 0° Centigrade; this bar is constantly heated at one end ($\lambda = 0$) up to the temperature 100° Centigrade. This example is found in J. Mikusiński [5], p. 229.

We have the following physical constants of silver:

| | |
|---|---|
| specific heat | $c = 0.055$ |
| conductivity | $k = 1.01$ |
| density | $\rho = 10.5$ |

and so

$$\alpha = \sqrt{0.055 \times 10.5/1.01} = 0.756 \quad \text{(approximately)}.$$

Hence

$$z(\lambda,t) = 100 \frac{2}{\sqrt{\pi}} \int_{0.378\lambda/\sqrt{t}}^{\infty} e^{-\nu^2} d\nu.$$

## §43. A BAR WITHOUT AN OUTGOING FLOW OF HEAT

We shall solve the equation

$$z_{\lambda\lambda}(\lambda,t) = \alpha^2 z_t(\lambda,t) \quad (0 \leqq \lambda \leqq \lambda_0,\ 0 \leqq t < \infty) \tag{43.1}$$

subject to the conditions

$$\begin{cases} z(\lambda,0) = 0 \quad (0 \leqq \lambda \leqq \lambda_0) \\ z(0,t) = v_1(t), \quad z_\lambda(\lambda_0,t) = 0 \quad (0 \leqq t < \infty). \end{cases} \tag{43.2}$$

Here $z_\lambda(\lambda_0,t)$ denotes the amount of heat flowing through the point $\lambda_0$ at the instant $t$. Thus the assumption $z_\lambda(\lambda_0,t) = 0$ means that the right end of the bar, like the whole of its lateral surface, is perfectly insulated from its environment.

The general solution of (43.1) is

$$z(\lambda) = ae^{-\alpha\lambda\sqrt{s}} + be^{\alpha\lambda\sqrt{s}} \quad (a,b \in C/C);$$

and $a$ and $b$ shall be determined from the boundary condition

$$\begin{cases} z(0) = a+b = v_1, \\ z_\lambda(\lambda_0) = -a\alpha\sqrt{s}\, e^{-\alpha\lambda_0\sqrt{s}} + b\alpha\sqrt{s}\, e^{\alpha\lambda_0\sqrt{s}} = 0. \end{cases}$$

That is,

$$a = \frac{v_1}{I+e^{-2\alpha\lambda_0\sqrt{s}}}, \quad b = \frac{v_1 e^{-2\alpha\lambda_0\sqrt{s}}}{I+e^{-2\alpha\lambda_0\sqrt{s}}}.$$

Hence we obtain

$$z(\lambda) = \frac{e^{-\alpha\lambda\sqrt{s}} + e^{-\alpha(2\lambda_0-\lambda)\sqrt{s}}}{I+e^{-2\alpha\lambda_0\sqrt{s}}} v_1.$$

Therefore, by using

$$\frac{I}{I+e^{-2\alpha\lambda_0\sqrt{s}}} = I + \sum_{\nu=1}^{\infty} (-1)^{\nu} e^{-2\nu\alpha\lambda_0\sqrt{s}},$$

we obtain the solution $z(\lambda)$, expanded into a series. The formula obtained in this way will differ from (41.8) only in signs of the terms of the infinite series. We shall not write down the details.

## §44. THE TEMPERATURE IN A BAR WITH A GIVEN INITIAL TEMPERATURE

FOURIER'S METHOD. Imagine that the temperature distribution in a bar of length $\lambda_0$ is given at the initial instant $t = 0$ by

$$z(\lambda,0) = \phi(\lambda) \qquad (0 \leqq \lambda \leqq \lambda_0)$$

so that

$$z_t(\lambda,t) = sz(\lambda,t) - [z(\lambda,0)] = sz(\lambda,t) - \phi(\lambda),$$

where $\phi(\lambda)$ is the *scalar multiplication operator*.

Thus we are concerned with solving the hyperfunction equation

$$z''(\lambda) - \alpha^2 sz(\lambda) = -\alpha^2\phi(\lambda) \qquad (0 \leqq \lambda \leqq \lambda_0) \tag{44.1}$$

under the boundary condition

$$z(0) = 0, \quad z(\lambda_0) = 0. \tag{44.2}$$

REMARK 44.1. In order to apply Proposition 29 to obtain a (special) hyperfunction solution of (44.1), we shall extend the *numerical-valued continuous function* $\phi(\lambda)$ to a numerical-valued continuous function defined for $\lambda \leqq 0$ also.

EXAMPLE 44.1. We take the case

$$\phi(\lambda) = \sin\frac{n\pi\lambda}{\lambda_0} \qquad (0 \leqq \lambda \leqq \lambda_0), \tag{44.3}$$

where $\phi(\lambda)$ is naturally extended to $\lambda \leqq 0$ also. Then the equation is

$$z''(\lambda) - \alpha^2 sz(\lambda) = -\alpha^2 \sin\frac{n\pi\lambda}{\lambda_0} \qquad (\alpha > 0). \tag{44.4}$$

Since the right hand term vanishes for $\lambda = 0$ and $\lambda = \lambda_0$, we shall tentatively substitute

$$z(\lambda) = a \sin\frac{n\pi\lambda}{\lambda_0} \qquad (a \in C/C) \tag{44.5}$$

for $z(\lambda)$ in (44.4). Then

$$-a\,\frac{n^2\pi^2}{\lambda_0^2} - a\alpha^2 s = -\alpha^2$$

so that we obtain

$$a = \frac{I}{s+n^2\beta^2} = \{\exp(-n^2\beta^2 t)\} \qquad (\beta = \pi/\alpha\lambda_0).$$

Hence we have a solution of (44.3) satisfying (44.2):

$$z(\lambda) = \{z(\lambda,t)\} = \left\{\exp(-n^2\beta^2 t)\,\sin\frac{n\pi\lambda}{\lambda_0}\right\} \tag{44.6}$$

PROPOSITION 38. The solution of (44.2)-(44.3) is uniquely determined by (44.6).

PROOF: Assume that there exists another solution $\hat{z}(\lambda)$ of (44.2)-(44.3). Then

$$w(\lambda) = z(\lambda) - \hat{z}(\lambda),$$

where $z(\lambda)$ is given by (44.6), satisfies

$$w''(\lambda) - \alpha^2 s w(\lambda) = 0, \quad w(0) = 0, \quad w(\lambda_0) = 0.$$

To prove that this $w(\lambda)$ is 0, we put

$$w(\lambda) = ae^{-\alpha\lambda\sqrt{s}} + be^{\alpha\lambda\sqrt{s}} \qquad (a,b \in C/C)$$

and choose $a$ and $b$ in such a way that $w$ satisfies the boundary condition

$$w(0) = a+b = 0, \quad w(\lambda_0) = ae^{-\alpha\lambda_0\sqrt{s}} + be^{\alpha\lambda_0\sqrt{s}} = 0.$$

Then $a = b = 0$, because the determinant

$$\begin{vmatrix} 1 & 1 \\ e^{-\alpha\lambda_0\sqrt{s}} & e^{\alpha\lambda_0\sqrt{s}} \end{vmatrix} = e^{\alpha\lambda_0\sqrt{s}} - e^{-\alpha\lambda_0\sqrt{s}}$$

is not zero. For, if $e^{\alpha\lambda_0\sqrt{s}} = e^{-\alpha\lambda_0\sqrt{s}}$, then

$$I = e^{-2\alpha\lambda_0\sqrt{s}}$$

or, more generally,

$$I = I^n = e^{-2\alpha\lambda_0 n\, s} \qquad (n = 1,2,3,\ldots). \tag{44.7}$$

This contradicts

$$e^{-\lambda\sqrt{s}} \leqq \left[\frac{3}{\lambda^2}\sqrt{(6/\pi e^3)}\right]h. \tag{44.8}$$

Indeed, for any $f(t) \in C[0,\infty)$, we have, by (44.7)-(44.8),

$$|f(t)| = \left|e^{-2\alpha\lambda_0 n\sqrt{s}}\{f(t)\}\right|$$

$$\leqq \left\{3\sqrt{(6/\pi e^3)}\,\frac{1}{(2\alpha\lambda_0 n)^2}\int_0^t |f(u)|\,du\right\} \quad (n = 1,2,\ldots).$$

This is absurd, as is seen by letting $n \to \infty$.

A CONSEQUENCE OF EXAMPLE 44.1. Let $\phi(\lambda)$ be given by

$$\phi(\lambda) = \sum_{n=1}^{\infty} \mu_n \sin\frac{n\pi\lambda}{\lambda_0} \qquad (0 \leqq \lambda \leqq \lambda_0), \tag{44.9}$$

where

$$\sum_{n=1}^{\infty} |\mu_n|\left(\frac{n\pi}{\lambda_0}\right)^2 < +\infty. \tag{44.10}$$

Then the equation

$$\begin{cases} z''(\lambda) - \alpha^2 s z(\lambda) = -\alpha^2 \sum_{n=1}^{\infty} \mu_n \sin\frac{n\pi\lambda}{\lambda_0}, \\ z(0) = 0, \quad z(\lambda_0) = 0 \end{cases} \tag{44.11}$$

is solved by

$$z(\lambda) = \{z(\lambda,t)\} = \left\{\sum_{n=1}^{\infty} \mu_n \exp\left(-\frac{n^2\pi^2}{\alpha^2\lambda_0^2}t\right)\sin\frac{n\pi\lambda}{\lambda_0}\right\}. \tag{44.12}$$

PROOF: By virtue of (44.10) and

$$0 \leqq \exp\left(-\frac{n^2\pi^2}{\alpha^2\lambda_0^2}t\right) \leqq 1, \quad \left|\sin\frac{n\pi\lambda}{\lambda_0}\right| \leq 1, \quad \left|\cos\frac{n\pi\lambda}{\lambda_0}\right| \leq 1, \tag{44.13}$$

the right-hand side of (44.12) and its term-wise differentiations

$$\frac{\partial}{\partial\lambda}, \quad \frac{\partial^2}{\partial\lambda^2} \quad \text{and} \quad \frac{\partial}{\partial t}$$

are all uniformly convergent on $0 \leqq \lambda \leqq \lambda_0$, $0 \leqq t$.

Therefore, (44.12) and its differentiations $\partial/\partial\lambda$, $\partial^2/\partial\lambda^2$ and $\partial/\partial t$ are obtained by term-wise differentiations of the right-hand side of (44.12). Hence, by Example 44.1, we easily prove that (44.12) is a solution of (44.11). Moreover, the uniqueness of this solution (44.12) is proved similarly as in Proposition 38.

REMARK 44.2. The solution of the heat equation by the above *Fourier's method* is customary in many textbooks, so we shall give only one other example of the method, in the next section.

REMARK 44.3. For the equation (44.1)-(44.2), we may obtain a hyperfunction solution by making use of Proposition 29 as it was stated in Remark 44.1. However, it would sometimes be difficult to prove the *smoothness* (or the differentiability) of the hyperfunction solution. For this reason, Fourier's method is sometimes easier to handle.

## §45. A HEAT-CONDUCTING RING

We shall discuss the temperature of a heat-conducting ring, given the temperature distribution at the initial instant $t = 0$. It will be assumed that for this ring there is no external flow of heat in or out.

If the length of the ring is $\lambda_0$, we can regard it in discussion as a bar of length $\lambda_0$ and consider the equation

$$z''(\lambda) - \alpha^2 s z(\lambda) = -\alpha^2\phi(\lambda), \tag{45.1}$$

where the function $\phi(\lambda)$, given in the interval $[0,\lambda_0]$ and thought of as *periodic with period* $\lambda_0$, represents the initial distribution of temperature.

We thus assume that $\phi(\lambda)$ is expressed as

$$\phi(\lambda) = \frac{\alpha_0}{2} + \sum_{n=1}^{\infty}\left(\alpha_n \cos\frac{2n\pi\lambda}{\lambda_0} + \beta_n \sin\frac{2n\pi\lambda}{\lambda_0}\right) \tag{45.2}$$

so that it will be natural to assume that the hyperfunction solution $z(\lambda)$ of (45.1)-(45.2) is given by

$$z(\lambda) = \frac{a_0}{2} + \sum_{n=1}^{\infty}\left(a_n \cos\frac{2n\pi\lambda}{\lambda_0} + b_n \sin\frac{2n\pi\lambda}{\lambda_0}\right), \tag{45.3}$$

where $a_n$ and $b_n \in \mathcal{C}/\mathcal{C}$ $(n = 0,1,2,\ldots)$.

In order to follow the reasoning of the preceding section, we shall assume that

$$\sum_{n=1}^{\infty} (|\alpha_n| + |\beta_n|)\left(\frac{2n\pi}{\lambda_0}\right)^2 < +\infty. \tag{45.4}$$

Then we obtain

$$a_n = \frac{\alpha_n}{s+4n^2\beta^2}, \quad b_n = \frac{\beta_n}{s+4n^2\beta^2} \qquad (\beta = \frac{\pi}{\alpha\lambda_0}) \tag{45.5}$$

so that

$$z(\lambda) = \{z(\lambda,t)\}$$

$$= \frac{\alpha_0}{2} + \sum_{n=1}^{\infty} \exp\left(-\frac{4n^2\pi^2}{\alpha^2\lambda_0^2} t\right)\left(\alpha_n \cos \frac{2n\pi\lambda}{\lambda_0} + \beta_n \sin \frac{2n\pi\lambda}{\lambda_0}\right). \tag{45.6}$$

PROPOSITION 39. (45.6) is the uniquely determined solution of (45.1) which is periodic in $\lambda$ with period $\lambda_0$.

PROOF: Assume the contrary and let $\hat{z}(\lambda) \neq \{z(\lambda,t)\}$ be another solution of (45.1) which is periodic with period $\lambda_0$.

Then

$$\tilde{z}(\lambda) = z(\lambda) - \hat{z}(\lambda) \neq 0$$

is a solution of

$$\tilde{z}''(\lambda) = \alpha^2 s\tilde{z}(\lambda) \tag{45.7}$$

and satisfies

$$\tilde{z}(\lambda+n\lambda_0) = \tilde{z}(\lambda) \qquad (n = \pm1,\pm2,\ldots).$$

As a solution of (45.7), we must have

$$\tilde{z}(\lambda) = ae^{-\alpha\lambda\sqrt{s}} + be^{\alpha\lambda\sqrt{s}} \qquad (a,b \in C/C).$$

By the periodicity of $\tilde{z}(\lambda)$, we have

$$\tilde{z}(0) = \tilde{z}(\lambda_0) = \tilde{z}(2\lambda_0),$$

i.e.,

$$\tilde{z}(0) = a + b = \tilde{z}(\lambda_0) = ae^{-\alpha\lambda_0\sqrt{s}} + be^{\alpha\lambda_0\sqrt{s}}, \tag{45.8}$$

and

$$\tilde{z}(0) = a + b = z(2\lambda_0) = ae^{-2\alpha\lambda_0\sqrt{s}} + be^{2\alpha\lambda_0\sqrt{s}} \tag{45.9}$$

hold.

Since $\tilde{z}(\lambda) \neq 0$, we must not have $a = b = 0$. Let $a \neq 0$. Then, from (45.8),

$$\frac{b}{a} = \frac{I - e^{-\alpha\lambda_0\sqrt{s}}}{e^{\alpha\lambda_0\sqrt{s}} - I},$$

because we have

$$e^{\gamma\sqrt{s}} \neq I \quad \text{for} \quad \gamma > 0 \quad \text{and for} \quad \gamma < 0. \tag{45.10}$$

(The proof of (45.10) is as follows: If $\gamma > 0$, then $e^{\gamma\sqrt{s}} = I$ implies

$$I = e^{-\gamma\sqrt{s}} = \left\{\frac{\gamma}{2\sqrt{\pi t^3}} \exp(-\gamma^2/4t)\right\} \in C$$

(see (27.5)), contrary to the Proposition 5 (§4).)

Now, as above, we have, from (45.9),

$$\frac{b}{a} = \frac{I - e^{-2\alpha\lambda_0\sqrt{s}}}{e^{2\alpha\lambda_0\sqrt{s}} - I}$$

and hence

$$\frac{I - e^{-\alpha\lambda_0\sqrt{s}}}{e^{\alpha\lambda_0\sqrt{s}} - I} = \frac{I - e^{-2\alpha\lambda_0\sqrt{s}}}{e^{2\alpha\lambda_0\sqrt{s}} - I} = \frac{I - e^{-\alpha\lambda_0\sqrt{s}}}{e^{\alpha\lambda_0\sqrt{s}} - I}\,\frac{(I + e^{-\alpha\lambda_0\sqrt{s}})}{(e^{\alpha\lambda_0\sqrt{s}} + I)},$$

proving, via (45.10), that

$$I = \frac{I + e^{-\alpha\lambda_0\sqrt{s}}}{e^{\alpha\lambda_0\sqrt{s}} + I} \quad \text{and so} \quad e^{\alpha\lambda_0\sqrt{s}} = e^{-\alpha\lambda_0\sqrt{s}}.$$

This implies

$$I = e^{-2\alpha\lambda_0\sqrt{s}},$$

contrary to (45.10).*

---

*The above "uniqueness proof" may be compared with that given by J. Mikusiński [5], p. 245-246.

## §46. NON-INSULATED HEAT CONDUCTION

So far, we have assumed that the bar is perfectly insulated from the environment. In this section, we assume that the heat loss at each point of the bar is proportional to the temperature difference at each point from the temperature of the environment, assumed to be constant at 0° Centigrade. We then have

$$z_{\lambda\lambda}(\lambda,t) - \alpha^2 z_t(\lambda,t) - \beta z(\lambda,t) = 0 \quad (\alpha > 0, \ \beta > 0).$$

Thus, assuming that the temperature of the bar at the initial instant $t = 0$ is 0° Centigrade, we have the equation

$$z''(\lambda) - (\alpha^2 s+\beta) z(\lambda) = 0 \quad (0 \leqq \lambda \leqq \lambda_0) \tag{46.1}$$

with the boundary condition

$$z(0) = v, \quad z(\lambda_0) = 0. \tag{46.2}$$

This equation is the same as the telegraph equation we have studied in §38, the cable without self-induction, so we do not go into detail.

However, for the convenience of the reader, we shall consider the case of an infinitely long bar for which $v = \{\omega\} = \frac{(\omega)}{s}$ :

$$z(\lambda) = \frac{\omega}{s} \exp(-\lambda \sqrt{\alpha^2 s+\beta}); \tag{46.3}$$

that is,

$$\begin{aligned} z(\lambda,t) = \frac{\omega}{2}\Big\{ & e^{-\lambda\sqrt{\beta}} \operatorname{Erf}\left(\frac{\sqrt{\beta t}}{\alpha} - \frac{\alpha\lambda}{2\sqrt{t}}\right) \\ & - e^{\lambda\sqrt{\beta}} \operatorname{Erf}\left(\frac{\sqrt{\beta t}}{\alpha} + \frac{\alpha\lambda}{2\sqrt{t}}\right) + 2 \cos h(\lambda\sqrt{\beta})\Big\}. \end{aligned} \tag{46.4}$$

# Answers to Exercises

§2.

(α) $\{t\}$

(β) $\{t^2/2\}$

(γ) $\{-4 \sin t - t\}$

(δ) $\{2 \sin t - 2 \cos t + 2\}$

§9.

1. $$\lim_{z\to 1} \frac{1}{(z+1)(z-2)} = \frac{-1}{2}, \quad \lim_{z\to 1} \left\{\frac{(z-1)}{(z-1)^2(z+1)(z-2)} - \frac{-1\cdot(z-1)}{2(z-1)^2}\right\}$$

$$= \lim_{z\to 1} \frac{z(z-1)}{2(z-1)(z+1)(z-2)} = \frac{-1}{4}, \quad \lim_{z\to -1} \frac{1}{(z-1)^2(z-2)} = \frac{-1}{12},$$

$$\lim_{z\to 2} \frac{1}{(z-1)^2(z+1)} = \frac{1}{3}$$

Hence $$\frac{I}{(s-1)^2(s+1)(s-2)} = \frac{1}{12}\left[\frac{-6}{(s-1)^2} + \frac{-3}{s-1} + \frac{-1}{s+1} + \frac{4}{s-2}\right].$$

2. $$\frac{I}{(s-\alpha)^2-\beta^2} = \frac{-1}{2\beta}\left(\frac{I}{(s-\alpha)+\beta} - \frac{I}{(s-\alpha)-\beta}\right)$$

3. $$\frac{s-\alpha}{(s-\alpha)^2-\beta^2} = \frac{1}{2}\left(\frac{I}{s-\alpha+\beta} + \frac{I}{s-\alpha-\beta}\right)$$

§10.

1. $e^{-t}(t^2 - 2t + 1)$

2. $y(t) = (t-1)^2 e^{-t}$

3. $y(t) = -\frac{1}{2} + \frac{1}{10} e^{2t} + \frac{2}{5} \cos t - \frac{1}{5} \sin t$

4. $y(t) = -\frac{1}{2} + e^{t} - \frac{11}{34}e^{4t} - \frac{3}{17}\cos t + \frac{5}{17}\sin t,$

$z(t) = -\frac{2}{3}e^{t} + \frac{22}{51}e^{4t} + \frac{4}{17}\cos t - \frac{1}{17}\sin t$

5. $x(t) = 2 - e^{t}, \quad y(t) = -2 + 4e^{t} - te^{t}, \quad z(t) = -2 + 5e^{t} - te^{t}$

6. $x(t) = \frac{28}{9}e^{3t} - e^{-t} - \frac{1}{9} - \frac{t}{3}, \quad y(t) = \frac{28}{9}e^{3t} + e^{-t} - \frac{1}{9} - \frac{t}{3}$

7. $x(t) = 2e^{2t} - 2\cos t - 3t\sin t$

$y(t) = \sin t + t\cos t - 2t\sin t$

§11.

1. Solvable if and only if $\cos 2\pi\alpha \neq 0$; and then

$$y(t) = \frac{1}{\alpha\cos 2\pi\alpha}\sin\alpha t.$$

2. $y(t) = \dfrac{e^{\alpha t} - e^{-\alpha t}}{\alpha(e^{2\pi\alpha} + e^{-2\pi\alpha})}$

§13.

$$\frac{1}{s^{1/2}(s-\alpha)} = h^{1/2}\{e^{\alpha t}\} = \frac{1}{\sqrt{\pi}}\int_0^t \frac{1}{\sqrt{t-u}}e^{\alpha u}\,du$$

$$= \frac{1}{\sqrt{\pi}}\int_0^t e^{\alpha(t-u)}\frac{1}{\sqrt{u}}\,du = \frac{e^{\alpha t}}{\sqrt{\pi}}\int_0^t e^{-\alpha u}\frac{1}{\sqrt{u}}\,du$$

$$= \frac{2e^{\alpha t}}{\sqrt{\pi}}\int_0^{\sqrt{t}} e^{-\alpha\lambda^2}\,d\lambda = \frac{2e^{\alpha t}}{\sqrt{\pi\alpha}}\int_0^{\sqrt{\alpha t}} e^{-x^2}\,dx = \frac{e^{\alpha t}}{\sqrt{\alpha}}\,\frac{2}{\sqrt{\pi}}\int_0^{\sqrt{\alpha t}} e^{-x^2}\,dx$$

$$= \frac{e^{\alpha t}}{\sqrt{\alpha}}\,\mathrm{Erf}(\sqrt{\alpha t})$$

§16.

1. $$\frac{I}{s^2}e^{-\alpha/s} = \sum_{n=0}^{\infty}(-1)^n\frac{\alpha^n h^{n+2}}{n!} = \left\{\sum_{n=0}^{\infty}(-1)^n\frac{\alpha^n t^{n+1}}{n!(n+1)!}\right\}$$

$$= \left\{(\sqrt{t/\alpha})\sum_{n=0}^{\infty}(-1)^n\frac{(2\sqrt{\alpha t})^{2n+1}}{2^{2n+1}n!(n+1)!}\right\} = \left\{(\sqrt{t/\alpha})\,J_1(2\sqrt{\alpha t})\right\}$$

2. $$\frac{I}{\sqrt{s}}e^{\alpha/s} = \sum_{n=0}^{\infty}\frac{\alpha^n h^{n+1/2}}{n!} = \left\{\sum_{n=0}^{\infty}\frac{\alpha^n t^{n-1/2}}{n!\Gamma(n+1/2)}\right\}$$

$$= \left\{\sum_{n=0}^{\infty}\frac{\alpha^n t^{n-1/2}}{n(n-1)\cdots 1(n-1/2)(n-3/2)\cdots(1/2)\Gamma(1/2)}\right\}$$

$$= \left\{\sum_{n=0}^{\infty} \frac{2^{2n}\alpha^n t^{n-1/2}}{(2n)(2n-1)\cdots 1}\right\} = \left\{\frac{1}{\sqrt{\pi t}} \cosh\,(2\sqrt{\alpha t})\right\}$$

3. See (16.13).

§20.

(α) Since $e^{-\beta/s^m} = e^{-\beta h^m} = I + \sum_{k=1}^{\infty} (-\beta)^k \frac{h^{mk}}{k!}$,

$$De^{-\beta/s^m} = DI + \sum_{k=1}^{\infty} (-\beta)^k \frac{1}{k!} (Dh^{mk})$$

$$= 0 + \sum_{k=1}^{\infty} (-\beta)^k \frac{1}{k!} (-mk\ h^{mk+1})$$

$$= \sum_{k=1}^{\infty} (-\beta)^{k-1} \frac{h^{m(k-1)}}{(k-1)!}\ m\beta\ h^{m+1}$$

$$= e^{-\beta/s^m} \frac{m\beta}{s^{m+1}}$$

(β) $\frac{Dy}{y} = \frac{(-2+2b+2)s + c}{s^2 + 1/4} = \frac{2bs + c}{s^2 + 1/4}$ .

$$= \frac{b + ci}{s + i/2} + \frac{b - ci}{s - i/2},$$

so that

$$y = C(s + i/2)^{b+ci}(s - i/2)^{b-ci} \qquad (i = \sqrt{-1}).$$

§29.

1. $y_0(\lambda) = \frac{-6\lambda I}{s^3} + \frac{-\lambda^3 I}{s} = \{-3t^2\lambda-\lambda^3\}$

2. $y_0(\lambda) = \frac{-24I}{s^5} + \frac{-12\lambda^2 I}{s^3} + \frac{-\lambda^4 I}{s} = \{-t^4 - 6t^2\lambda^2 - \lambda^4\}$

§36.

(α) $\frac{1}{s^2+a^2} = \{\frac{1}{a} \sin at\} < \frac{1}{a}$

(β) $\frac{1}{\sqrt{s}} e^{-\lambda\sqrt{s}} = \left\{\frac{1}{\sqrt{t}} \exp(-\lambda^2/4t)\right\}$.

The function on the right-hand side takes its maximum value

$$\frac{1}{\lambda} \frac{2}{\sqrt{\pi e}} \quad \text{at} \quad t = \frac{\lambda^2}{2}.$$

# Formulas and Tables

## I. Special Functions

$$\Gamma(\lambda) = \int_0^\infty t^{\lambda-1} e^{-t} dt \quad (\operatorname{Re}\lambda > 0) \qquad (\S 12)$$

$$\Gamma(\lambda+1) = \lambda\Gamma(\lambda), \quad \Gamma(n) = (n-1)! \quad (n = 1,2,\cdots)$$

$$B(\lambda,\mu) = \frac{\Gamma(\lambda)\Gamma(\mu)}{\Gamma(\lambda+\mu)} \int_0^1 t^{\lambda-1}(1-t)^{\mu-1} dt \quad (\operatorname{Re}\lambda > 0, \operatorname{Re}\mu > 0) \qquad (\S 12)$$

$$\Gamma(1/2) = \sqrt{\pi} = 2\int_0^\infty e^{-v^2} dv$$

$$\operatorname{Erf} t = \frac{2}{\sqrt{\pi}} \int_0^t e^{-v^2} dv \qquad (\S 13)$$

$$\operatorname{Cerf} t = \frac{2}{\sqrt{\pi}} \int_t^\infty e^{-v^2} dv = 1 - \operatorname{Erf} t$$

$$J_n(t) = \sum_{k=0}^\infty (-1)^k \frac{t^{n+2k}}{2^{n+2k} k!(k+n)!} \quad (n = 0,1,2,\cdots) \qquad (\S 15, \S 20)$$

$$J_n(it) = i^n \sum_{k=0}^\infty \frac{t^{n+2k}}{2^{n+2k} k!(k+n)!} \quad (n = 0,1,2,\cdots)$$

$$L_n^\alpha(t) = \sum_{k=0}^\infty \binom{n+\alpha}{n-k} \frac{(-t)^k}{k!} \qquad (\S 20)$$

$(1+z)^\gamma = \exp(\gamma \log(1+z))$ for complex numbers $\gamma$ and $z$

$$= \sum_{k=0}^\infty \binom{\gamma}{k} z^k \quad (\text{convergent for } |z| < 1) \qquad (\S 14)$$

## II. Formulas of the Operational Calculus

$C = C[0,\infty)$ (§1)

$C_H = \{\frac{f}{h^k};\ f \in C \text{ and } k = 1,2,\ldots\}$ (§3)

$C/C$: convolution quotients $\frac{f}{g}$ $(f,g \in C$ with $g \neq 0)$ (§18)

$\{a(t)\} + \{b(t)\} = \{a(t) + b(t)\}$,

$\{a(t)\}\{b(t)\} = \left\{\int_0^t a(t-u)b(u)\,du\right\}$,

$\alpha\{f(t)\} = \{\alpha f(t)\}$

$h\{f(t)\} = \left\{\int_0^t f(u)\,du\right\}$

$s\{f(t)\} = \{f'(t)\} + [f(0)],\ [f(0)] = s\{f(0)\}$ (§5)

$s^n f = f^{(n)} + s^{n-1}[f(0)] + \cdots + s[f^{(n-2)}(0)] + [f^{(n-1)}(0)]$

$T^\alpha\{f(t)\} = \{e^{\alpha t}f(t)\},\quad T^\alpha R(s) = R(s-\alpha)$

$T^\alpha \frac{b}{a} = \frac{T^\alpha b}{T^\alpha a}$ (§20, §37)

$Df = \{-tf(t)\},\quad D(fg) = (Df)g + f(Dg)$

$D\frac{b}{a} = \frac{(Db)a - b(Da)}{a^2}$ (§19)

$\int_0^\infty e^{-\lambda s} f(\lambda)\,d\lambda = \{f(t)\}$ (§26)

$\frac{\partial}{\partial\lambda} e^{-\lambda s} = -se^{-\lambda s},\quad e^o = I$ (§23)

$\frac{\partial}{\partial\lambda} e^{-\lambda s^{1/2}} = -s^{1/2}\, e^{-\lambda s^{1/2}},\quad e^o = I$ (§26)

## III. Tables of Hyperfunctions $\subseteq C/C$

$\frac{I}{s} = h = \{1\},\quad I = \frac{f}{f}\ (f(t) \not\equiv 0)$ (§5)

$\frac{I}{s^n} = h^n = \left\{\frac{t^{n-1}}{(n-1)!}\right\}$ $(n = 1,2,\ldots)$ (§6)

$\frac{I}{s^\lambda} = h^\lambda = \left\{\frac{t^{\lambda-1}}{\Gamma(\lambda)}\right\}$ $(\mathrm{Re}\ \lambda > 0)$ (§13)

$$h^{\gamma} = \frac{h^{\gamma+n}}{h^n} = \frac{\Gamma(\gamma+n)^{-1}t^{\gamma+n-1}}{\Gamma(n)^{-1}t^{n-1}} \quad \text{for complex numbers } \gamma$$

$$(\text{integer } n \geqq 1 \text{ such that } \operatorname{Re}(\gamma+n) > 1) \qquad (\S 19)$$

$$Dh^{\gamma} = -\gamma h^{\gamma+1} \qquad (\S 19)$$

$$D(I-\alpha h)^{\gamma} = \gamma(I-\alpha h)^{\gamma-1}\alpha h^2, \text{ where}$$

$$(I-\alpha h)^{\gamma} = \sum_{k=0}^{\infty} \binom{\gamma}{k}(-\alpha)^k h^k \qquad (\S 19)$$

$$D(s-\alpha I)^{\gamma} = D\,\frac{(I-\alpha h)^{\gamma}}{h^{\gamma}} = \gamma(s-\alpha I)^{\gamma-1} \qquad (\S 19)$$

$$\frac{I}{\sqrt{s}} = h^{1/2} = \left\{\frac{1}{\sqrt{\pi t}}\right\} \qquad (\S 13)$$

$$\frac{I}{s-\alpha} = \{e^{\alpha t}\} \qquad (\S 6)$$

$$\frac{I}{(s-\alpha)^{\lambda}} = \left\{\frac{t^{\lambda-1}}{\Gamma(\lambda)}\, e^{\alpha t}\right\} \quad (\operatorname{Re}\lambda > 0) \qquad (\S 13)$$

$$\frac{I}{\sqrt{s+\alpha}} = \left\{\frac{1}{\sqrt{\pi t}}\, e^{-\alpha t}\right\} \qquad (\S 13)$$

$$\frac{I}{s\sqrt{s+\alpha}} = \left\{\frac{1}{\sqrt{\alpha}} \operatorname{Erf} \sqrt{\alpha t}\right\} \quad (\alpha > 0) \qquad (\S 13)$$

$$s^{\gamma} = h^{-\gamma}, \text{ in particular } s^0 = h^0 = I \text{ and } Ds = I \qquad (\S 19)$$

$$(s-\alpha I)^{\gamma}(s-\alpha I)^{\delta} \text{ for complex numbers } \gamma \text{ and } \delta$$

$$= (s-\alpha I)^{\gamma+\delta} \qquad (\S 19)$$

$$h^{\gamma}h^{\delta} = h^{\gamma+\delta} \qquad (\S 19)$$

$$\frac{I}{(s-\alpha)^2+\beta^2} = \left\{\frac{1}{\beta}\, e^{\alpha t} \sin \beta t\right\}$$

$$\frac{I}{[(s-\alpha)^2+\beta^2]^2} = \left\{\frac{e^{\alpha t}}{2\beta^2}\left[\frac{1}{\beta} \sin \beta t - t \cos \beta t\right]\right\}$$

$$\frac{s-\alpha}{(s-\alpha)^2+\beta^2} = \{e^{\alpha t} \cos \beta t\} \qquad (\S 9)$$

$$\frac{I}{(s-\alpha)^2-\beta^2} = \left\{\frac{1}{\beta}\, e^{\alpha t} \sinh \beta t\right\}$$

$$\frac{s-\alpha}{(s-\alpha)^2-\beta^2} = \{e^{\alpha t} \cosh \beta t\}$$

$$\frac{I}{(s^2+\alpha^2)^{1/2}} = \{J_0(\alpha t)\}$$

$$\frac{I}{(s^2-\alpha^2)^{1/2}} = \{J_0(i\alpha t)\}$$

$$\frac{(s^2+\alpha^2)^{1/2}-s}{(s^2+\alpha^2)^{1/2}} = \{\alpha J_1(\alpha t)\} \qquad (\S 16)$$

$$((s^2+\alpha^2)^{1/2}-s)^n = \left\{\frac{n\alpha^n}{t} J_n(\alpha t)\right\} \qquad (n = 1,2,\ldots)$$

$$\frac{((s^2+\alpha^2)^{1/2}-s)^n}{(s^2+\alpha^2)^{1/2}} = \{\alpha^n J_n(\alpha t)\} \qquad (n = 0,1,2,\ldots)$$

$$\frac{I}{s} e^{-\lambda/s} = \{J_0(2\sqrt{\lambda t})\}$$

$$\frac{I}{s^2} e^{-\lambda/s} = \{\sqrt{(t/\lambda)}\, J_1(2\sqrt{\lambda t})\} \qquad (\S 16)$$

$$\frac{I}{\sqrt{s}} e^{-\lambda/s} = \left\{\frac{1}{\sqrt{\pi t}} \cos 2\sqrt{\lambda t}\right\}$$

$$\frac{I}{\sqrt{s}} e^{\lambda s} = \left\{\frac{1}{\sqrt{\pi t}} \cosh(2\sqrt{\lambda t})\right\}$$

$$e^{-\lambda s} = s^2\{h_1(\lambda,t)\} = s^2\left\{\int_0^t H_\lambda(u)\,du\right\} = s^2\left\{\begin{matrix} 0, & 0 \leqq t < \lambda \\ t-\lambda, & 0 \leqq \lambda < t \end{matrix}\right\} \qquad (\S 22)$$

$$\exp(-\lambda\sqrt{s}) = \left\{\frac{\lambda}{2\sqrt{\pi t^3}} \exp(-\lambda^2/4t)\right\}$$

$$\frac{1}{\sqrt{s}} \exp(-\lambda\sqrt{s}) = \left\{\frac{1}{\sqrt{\pi t}} \exp(-\lambda^2/4t)\right\} \qquad (\S 27)$$

$$\frac{1}{s} \exp(-\lambda\sqrt{s}) = \{\mathrm{Cerf}(\lambda/2\sqrt{t})\}$$

$$\exp(\lambda[s-(s^2+\alpha^2)^{1/2}]) = I - \left\{\frac{\lambda}{\sqrt{t^2+2\lambda t}} \alpha J_1(\alpha\sqrt{t^2+2\lambda t})\right\}$$

$$\frac{\exp(\lambda[s-(s^2+\alpha^2)^{1/2}])}{(s^2+\alpha^2)^{1/2}} = \left\{J_0(\alpha\sqrt{t^2+2\lambda t})\right\} \qquad (\S 28)$$

$$\exp(\lambda[s-(s^2-\alpha^2)^{1/2}]) = I - \left\{\frac{\lambda}{\sqrt{t^2+2\lambda t}} i\alpha J_1(i\alpha\sqrt{t^2+2\lambda t})\right\}$$

$$\frac{\exp(\lambda[s-(s^2-\alpha^2)^{1/2}])}{(s^2-\alpha^2)^{1/2}} = \left\{J_0(i\alpha\sqrt{t^2+2\lambda t})\right\}$$

$$\exp(\lambda[s-(s^2+2\alpha s)^{1/2}]) = e^{-\alpha\lambda} - \frac{\lambda}{\sqrt{t^2+2\lambda t}} e^{-\alpha(\lambda+t)} i\alpha J_1(i\alpha\sqrt{t^2+2\lambda t})$$

(§37)

$$\frac{\exp(\lambda[s-(s^2+2\alpha s)^{1/2}])}{(s^2+2\alpha s)^{1/2}} = e^{-\alpha(\lambda+t)} J_0(i\alpha\sqrt{\lambda^2+2\lambda t})$$

$$\exp(-\lambda(s^2+\alpha^2)^{1/2}) = e^{-\lambda s} - \left\{\begin{matrix} 0, & 0 \leqq t < \lambda \\ \frac{\lambda}{\sqrt{t^2-\lambda^2}}\alpha J_1(\alpha\sqrt{t^2-\lambda^2}), & 0 \leqq \lambda < t \end{matrix}\right\}$$

$$\frac{\exp(-\lambda(s^2+\alpha^2)^{1/2})}{(s^2+\alpha^2)^{1/2}} = \left\{\begin{matrix} 0, & 0 \leqq t < \lambda \\ J_0(\alpha\sqrt{t^2-\lambda^2}), & 0 \leqq \lambda < t \end{matrix}\right\}$$

(§28)

$$\exp(-\lambda(s^2-\alpha^2)^{1/2}) = e^{-\lambda s} - \left\{\begin{matrix} 0, & 0 \leqq t < \lambda \\ \frac{\lambda}{\sqrt{t^2-\lambda^2}} i\alpha J_1(i\alpha\sqrt{t^2-\lambda^2}), & 0 \leqq \lambda < t \end{matrix}\right\}$$

$$\frac{\exp(-\lambda(s^2-\alpha^2)^{1/2})}{(s^2-\alpha^2)^{1/2}} = \left\{\begin{matrix} 0, & 0 \leqq t < \lambda \\ J_0(i\alpha\sqrt{t^2-\lambda^2}), & 0 \leqq \lambda < t \end{matrix}\right\}$$

$$\exp(-\lambda(s^2+2\alpha s)^{1/2}) = e^{-\alpha\lambda}e^{-\lambda s} - \left\{\begin{matrix} 0, & 0 \leqq t < \lambda \\ \frac{\lambda}{\sqrt{t^2-\lambda^2}} e^{-\alpha t} i\alpha J_1(i\alpha\sqrt{t^2-\lambda^2}), & 0 \leqq \lambda < t \end{matrix}\right\}$$

$$\frac{\exp(-\lambda(s^2+2\alpha s)^{1/2})}{(s^2+2\alpha s)^{1/2}} = \left\{\begin{matrix} 0, & 0 \leqq t < \lambda \\ e^{-\alpha t} J_0(i\alpha\sqrt{t^2-\lambda^2}), & 0 \leqq \lambda < t \end{matrix}\right\}$$

$$\exp(\lambda[s-((s-\alpha)^2+\beta^2)^{1/2}]) = e^{\alpha\lambda} - \left\{\frac{\lambda}{\sqrt{t^2+2\lambda t}} e^{\alpha(\lambda+t)} \beta J_1(\beta\sqrt{t^2+2\lambda t})\right\} \quad (§37)$$

$$\frac{\exp(\lambda[s-((s-\alpha)^2+\beta^2)^{1/2}])}{((s-\alpha)^2+\beta^2)^{1/2}} = \left\{e^{\alpha(\lambda+t)} J_0(\beta\sqrt{t^2+2\lambda t})\right\}$$

$$\exp(-\lambda[(s-\alpha)^2+\beta^2]^{1/2}) = e^{\alpha\lambda}e^{-\lambda s} - \left\{\begin{matrix} 0, & 0 \leqq t < \lambda \\ \frac{\lambda}{\sqrt{t^2-\lambda^2}} e^{\alpha t} \beta J_1(\beta\sqrt{t^2-\lambda^2}), & 0 \leqq \lambda < t \end{matrix}\right\}$$

$$\frac{\exp(-\lambda[(s-\alpha)^2+\beta^2]^{1/2})}{((s-\alpha)^2+\beta^2)^{1/2}} = \left\{\begin{matrix} 0, & 0 \leqq t < \lambda \\ e^{\alpha t} J_0(\beta\sqrt{t^2-\lambda^2}, & 0 \leqq \lambda < t \end{matrix}\right\}$$

(There are also those obtained by substituting $\sqrt{-1}\,\beta$ for $\beta$ in the last two formulas above.)

$$e^{-\lambda\sqrt{s}} \leqq \left[\frac{3}{\lambda^2}\,(\sqrt{6/\pi e^3}\right]h \qquad (\lambda > 0)$$

(§36)

$$\frac{1}{\sqrt{s}}\,e^{-\lambda\sqrt{s}} \leqq \left[(\sqrt{2/\pi e})\,\frac{1}{\lambda}\right]h \qquad (\lambda > 0)$$

# References

1. Berg, E. J.: Heaviside's Operational Calculus as Applied to Engineering and Physics, McGraw-Hill Book Company (1936).

2. Doetsch, Gustav: Einführung in Theorie und Anwendung der Laplace-Transformation, Birkhäuser Verlag (1958).

3. Erdélyi, Arthur: Operational Calculus and Generalized Functions, Holt, Reinhart, and Winston (1962).

4. Heaviside, Oliver: Electromagnetic Theory, I-III, London (1893-1899).

5. Mikusiński, Jan: Operational Calculus, Pergamon Press (1959).

6. Mikusiński, Jan: The Bochner Integral, Academic Press (1978).

7. Krabbe, Gregers: Operational Calculus, Springer-Verlag (1970).

8. Okamoto, Shuichi: A simplied derivation of Mikusiński's operational calculus, Proc. Jap Acad. 50, Ser. A, No. 1 (1979), 1-5.

9. Yosida, Kôsaku-Okamoto, Shûichi: A note on Mikusiński's operational calculus, Proc. Jap. Acad. 56, Ser. A, No. 1 (1980), 1-3.

10. Yosida, Kôsaku: The algebraic derivative and Laplace's differential equation, Proc. Jap. Acad. 59, Ser. A, No. 1 (1983), 1-4.

11. Yosida, Kôsaku-Matsuura, Shigetake: A note on Mikusiński's proof of the Titchmarsh convolution theorem, to be published in the Contemporary Mathematics Series of the Amer. Math. Soc.

# Propositions and Theorems in Sections

| Proposition | Section | Theorem | Section |
|---|---|---|---|
| 1 | 1 | 1 | 1 |
| 2 | 2 | 2 | 3 |
| 3 | 3 | 3 | 5 |
| 4 | 3 | 4 | 5 |
| 5 | 4 | 5 | 6 |
| 6 | 7 | 6 | 7 |
| 7 | 8 | 7 | 8 |
| 8 | 9 | 8 | 8 |
| 9,9' | 13,13 | 9,9' | 9,9 |
| 10 | 13 | 10 | 12 |
| 11 | 15 | 11 | 13 |
| 12 | 18 | 12 | 14 |
| 13 | 19 | 13 | 15 |
| 14 | 19 | 14 | 16 |
| 15 | 19 | 15 | 16 |
| 16 | 20 | 16 | 17 |
| 17,17' | 22,22 | 17 | 19 |
| 18 | 23 | 18 | 20 |
| 19 | 23 | 19 | 21 |
| 20 | 24 | 20 | 21 |
| 21 | 25 | 21 | 24 |
| 22 | 27 | 22 | 26 |
| 23 | 27 | 23 | 27 |
| 24 | 28 | 24 | 29 |
| 25 | 28 | | |
| 26 | 28 | | |
| 27 | 28 | | |
| 28 | 29 | | |
| 29 | 29 | | |
| 30 | 30 | | |
| 31 | 31 | | |
| 32 | 36 | | |

| Proposition | Section |
|---|---|
| 33 | 36 |
| 34 | 36 |
| 35 | 36 |
| 36 | 39 |
| 37 | 42 |
| 38 | 44 |
| 39 | 45 |

# Index


Algebraic derivative, 53-59

Bessel differential equation, 61-63

Bessel functions, 40-46

Beta function, 32

Binomial theorem, 39

Boundary value problems, 29-31

Confluent hypergeometric differential equation, 63-64

Convolution, 1, 76

Convolution quotient, 12

Convolution ring, 3

Convolution theorem, 11-12, 47

D'Alembert method, 63, 118-123

Delta function, 78

Differentiation operator, 9, 76

Electric circuits, 28-29

Entire function, 33

Error function, 37

Euler's integrals, 32

Exponential hyperfunctions, 74-76, 83-86, 94-105

Fourier's method, 150, 152, 153

Fractional differentiation, 37

Fractional integration, 35

Fractional powers, 34-38

Gamma function, 32

Gauss differential equation, 64

General powers, 38

Generalized continuity, 81

Generalized derivative, 10, 79

Generalized functions, 8-9

Generalized integral, 92

Generalized Laplace integral, 93

Heat equation, 106, 145-156

Heaviside's unit function, 74

Hyperfunction, 8-9, 50-51

Impulsive force, 77-78

Integration, 3

Inverse (see "multiplicative inverse")

Initial conditions, 14, 108

Initial value problem for ODEs, 14, 22

Laguerre polynomials, 65-66

Laguerre differential equation, 64-66

Laplace transform, 25, 53

Logarithmic hyperfunction, 99, 102-105

Moments, Mikusiński's theorem of, 70, 72-73, 89
Multiplicative inverse, 13, 52, 84
Multiplicative unit, 7, 51

Partial fraction decomposition, 18

Rational functions, 17-18

Scalar multiplication, 9-10
Shift operator, 74-76, 83-86

Telegraph equation, 106, 124-144
Titchmarsh convolution theorem, 12, 47
Thomson cable, 128

Uniqueness of solution, 23, 84, 98, 99, 108, 114, 120, 129, 146, 147, 151, 154

Wave equation, 106, 108-123
Weierstrass' theorem, 70-71

Zero factors, 15-16

# Applied Mathematical Sciences

*cont. from page ii*

36. Bengtsson/Ghil/Källén: **Dynamic Meterology: Data Assimilation Methods.**
37. Saperstone: **Semidynamical Systems in Infinite Dimensional Spaces.**
38. Lichtenberg/Lieberman: **Regular and Stochastic Motion.**
39. Piccinini/Stampacchia/Vidossich: **Ordinary Differential Equations in $R^n$.**
40. Naylor/Sell: **Linear Operator Theory in Engineering and Science.**
41. Sparrow: **The Lorenz Equations: Bifurcations, Chaos, and Strange Attractors.**
42. Guckenheimer/Holmes: **Nonlinear Oscillations, Dynamical Systems and Bifurcations of Vector Fields.**
43. Ockendon/Tayler: **Inviscid Fluid Flows.**
44. Pazy: **Semigroups of Linear Operators and Applications to Partial Differential Equations.**
45. Glashoff/Gustafson: **Linear Optimization and Approximation: An Introduction to the Theoretical Analysis and Numerical Treatment of Semi-Infinite Programs.**
46. Wilcox: **Scattering Theory for Diffraction Gratings.**
47. Hale et al.: **An Introduction to Infinite Dimensional Dynamical Systems — Geometric Theory.**
48. Murray: **Asymptotic Analysis.**
49. Ladyzhenskaya: **The Boundary-Value Problems of Mathematical Physics.**
50. Wilcox: **Sound Propagation in Stratified Fluids.**
51. Golubitsky/Schaeffer: **Bifurcation and Groups in Bifurcation Theory, Vol. I.**
52. Chipot: **Variational Inequalities and Flow in Porous Media.**
53. Majda: **Compressible Fluid Flow and Systems of Conservation Laws in Several Space Variables.**
54. Wasow: **Linear Turning Point Theory.**
55. Yosida: **Operational Calculus: A Theory of Hyperfunctions.**